Laser-Arc Processes and their Applications in Welding and Material Treatment

Welding and Allied Processes
A series of books and monographs on welding and other processes of metal treatment
Edited by B.E. Paton, E.O. Paton Electric Welding Institute, Kiev, Ukraine and P. Seyffarth,
Welding Training and Research Institute, Rostock, Germany

Volume 1

Laser-Arc Processes and their Applications in Welding and Material Treatment
P. Seyffarth and I.V. Krivtsun

This book is part of a series. The publisher will accept continuation orders which may be cancelled at
any time and which provide for automatic billing and shipping of each title in the series upon
publication. Please write for details.

Laser-Arc Processes and their Applications in Welding and Material Treatment

Edited by

Peter Seyffarth
Welding Training and Research Institute
Mecklenburg-Vorpommern GmbH
Rostock, Germany

and

Igor V. Krivtsun
E.O. Paton Electric Welding institute
National Academy of Sciences of Ukraine
Kiev, Ukraine

CRC Press
Taylor & Francis Group
Boca Raton London New York

CRC Press is an imprint of the
Taylor & Francis Group, an **informa** business
A TAYLOR & FRANCIS BOOK

First published 2002 by Taylor & Francis

CRC Press
Taylor & Francis Group
6000 Broken Sound Parkway NW, Suite 300
Boca Raton, FL 3487-2742

© 2002 by Taylor & Francis Group, LLC
CRC Press is an imprint of Taylor & Francis Group, an Informa business

No claim to original U.S. Government works

ISBN-13: 978-0-415-26961-2 (hbk)

This book has been produced from camera-ready copy supplied by the authors

Visit the Taylor & Francis Web site at
http://www.taylorandfrancis.com

and the CRC Press Web site at
http://www.crcpress.com

British Library Cataloguing in Publication Data
A catalogue record for this book is available from the British Library

Library of Congress Cataloging in Publication Data
A catalog record for this book has been requested

Contents

Introduction to the Series

The technology for permanently joining various materials by welding is among the most outstanding achievements of the end of nineteenth – beginning of twentieth centuries, and today most industries use welding technology.

Much work is underway on upgrading welding processes, widening their fields of application and improving the quality of welded structures. Further development of classical welding methods led to the emergence of a large number of new technologies for welding, thermal cutting, special electrometallurgy, surfacing, spraying and surface treatment, which have useful commercial applications. Nowadays welding is employed on earth, under water and in space.

Extensive R&D efforts are in progress throughout the world in fields such as metallurgy of welding, strength of welded structures, mechanization and automation of welding processes and many other welding-related areas. Welding research centres with a high scientific potential have been established.

There is an intensive exchange of scientific information between the research centres, although language barriers and other factors hamper this exchange. In particular, in the territory of the former Soviet Union, with numerous teams of scientists successfully working in the field of welding science and technology and possessing unique welding technologies, results of their work have been published until recently mostly in Russian or have been classified. This made them unavailable to specialists in Western countries.

The purpose of the new series of books published in the English language is to introduce the scientific community to the latest results of R&D in the field of welding and related technologies, conducted mostly at the E. O. Paton Electric Welding Institute of the National Academy of Sciences of Ukraine in collaboration with specialists from other countries.

The series is intended to fill the gap and provide specialists with results proven in practice, which can lead to new engineering solutions.

B. Paton,
P. Seyffarth

Preface

At present, alongside the conventional methods of material joining and treatment that use arc (plasma), laser or electron beam heating of materials, hybrid or combined processes are under development. The essence of these processes is in the simultaneous use of two different heat sources, for example, laser beam and electric arc. One of the main reasons for the appearance of hybrid laser-arc methods of welding, cutting and surface treatment of metals is the desire to combine the advantages of the established arc and laser technologies and to overcome their limitations.

The first investigations of the combined laser-arc welding and treatment of materials showed that they have many peculiarities, which cannot be explained by a simple superposition of the properties of the heat sources used taken separately. For example, it was established that the simultaneous laser-arc effect on metals increases efficiency of utilisation of both laser and arc power and improves spatial stabilisation of the arc heating spot at the surface of a workpiece. Besides, if the laser beam passes through the arc plasma, its absorption in the plasma under certain conditions causes essential changes in the energy balance of the arc discharge. In the case when the power put into the arc by the laser radiation is commensurate with its electric power, a special type of discharge, i.e. the combined laser-arc discharge, is realised. Therefore, the further development of these hybrid technological processes requires a detailed study of the straight interaction of laser beam and electric arc, used in hybrid heat source, as well as an investigation of the features of its simultaneous effect on a workpiece.

Despite the increasing number of papers on laser-arc methods for metal processing, in current scientific and technical literature, the problem of direct interaction between focused laser radiation and arc discharge plasma is not given the consideration it is due. Reasoning from this, the aim of the present book, as set by the authors, is not only to provide a generalisation and systematisation of the known experimental data on the hybrid processes and the examples of their practical applications, but also to acquaint the readers with the results of the theoretical investigation and computer simulation of the laser-arc discharge. These results can help to reveal the prospects for rational practical application of the combined discharge and development on this basis of integrated laser-arc plasma torches for various technological processes.

The book is compiled in the following way. The introduction is written jointly by the authors, Chapters 1, 2, 3 are written by I.V. Krivtsun and Chapter 4—by P. Seyffarth. At the same time the authors have to acknowledge that obviously it would not have been possible to produce a book of this kind without the invaluable assistance of all those who collaborated with us. We owe a particular debt to Profs. V. S. Gvozdetsky and H. Wohlfarth; Drs. A. I. Som and H. Staufer; Mrs. E. Ya. Bossenko, Mr. K. Fronius and Mr. J. Hoffmann without whose advice, suggestions and permission to use the results of their research almost every page of this book would have less to offer.

It is with great pleasure that we express our personal thanks to Professor W. M. Steen who was kind enough to read and comment on the typescript and who has made an enormous contribution to the clarity, accuracy and completeness of the finished text.

The authors would like to express their gratitude to Dr. A. T. Zelnichenko for the idea of publishing the Welding and Allied Processes book series and for his efforts to realise the aim of this book.

We are of deeply grateful to Mrs. L. M. Yarinich for her help with compiling Chapter 1 and Mrs. T. K. Vassylenko for translating the book into English. We also would like to thank Miss T. Yu. Snegireva and Miss I. S. Batasheva for the clear illustrations and Miss N. V. Mnyshenko for preparation of CRC of the book.

Last but not least, we must thank those who are close and dear to us for their support and tolerance during the writing of this book, which has separated us from them for too many hours.

P. Seyffarth
I.V. Krivtsun
Rostock, June 2000

List of Notations and Symbols

A_R – Richardson constant

$A_{1\omega} \equiv A_\omega$, $A_{2\omega}$ – complex amplitudes of electric field of incident and reflected laser beam

b_α – mobility of charged particles of α kind

c – velocity of light

C_2, C_3 – equilibrium constants for the first- and second-order ionisation reactions

C_p – specific heat of plasma gas at constant pressure

d – distance from plasma torch exit section to anode (workpiece)

D_e, D_i – electron diffusion coefficient, coefficient of ion diffusion in slightly ionised plasma

$D_{T\alpha}$ – coefficient of thermal diffusion of particles of α kind

$D_{\alpha\beta}$ – coefficient of mutual diffusion of heavy particles of α and β kinds in two-temperature plasma

D_e^A – coefficient of ambipolar diffusion of electrons

$D_{T_e}^A$ – coefficient of ambipolar thermal diffusion of electrons

$[D_{\alpha\beta}]_1$ – coefficient of mutual diffusion of particles of α and β kinds for isothermal plasma in the first approximation by Chapmen-Enskog method

e – elementary charge (charge of electron)

$\mathbf{e}$ – unit vector of laser radiation polarisation

$\mathbf{E}$, E_r, $E_z \equiv E$ – electric field vector in plasma, its radial and axial components

E_c – electric field strength at cathode spot surface

$\mathbf{E}_\omega$ – electric field vector of laser radiation beam

f_α – distribution function of α kind particles

$\bar{f}$, f' – averaging according to Reynolds and corresponding pulsations

F – distance of initial beam focus from cathode tip (cathode nozzle exit)

g – Gaunt factor

$\tilde{g}$, g' – averaging according to Favre and corresponding pulsations

G, G_0, G_1, G_2 – mass flow rates of plasma gas

G^V, G_1^V – volume flow rates of plasma gas

h – specific enthalpy of plasma gas

$\hbar$ – Plank constant divided by 2π

$\mathbf{h}_\alpha$ – energy flow of α kind particles

H_m – Hermite polynomials

$H_\varphi \equiv H$ – azimuthal component of magnetic field in plasma

I – arc current

$\mathbf{j}$, j_r, $j_z \equiv j$ – total current density in plasma, its radial and axial components

$\mathbf{j}_e$, $\mathbf{j}_i$ – electron and ion components of current density in plasma

j_e, j_i – densities of electron and ion currents from plasma to cathode

j_{em} – density of emitted electron current

$\mathbf{j}_c, j_c, \langle j_c \rangle$ – current density in cathode body, its local and averaged values at cathode spot surface

$j_{ra}, j_{za} \equiv j_a$ – radial and axial components of current density within anode region

$k_z \equiv k$ – wave vector of laser radiation

k_B – Boltzmann constant

K – total pulse flow of plasma

K_3, K_4 – three-body recombination coefficients for singly and doubly charged ions

K_{la} – energy input coefficient for laser-arc process

l_1, l_2 – thicknesses of space charge layer and ionisation layer

l_m – mixing length

L – length of discharge calculation domain

L_C – length of plasma-shaping channel

m_e – electron mass

m_α – mass of the α particle

M, M_{Ar}, M_{He} – masses of atoms of plasma gas

M_0, M_1, M_2 – whirling moments of plasma gas

n_e, n_{ea} – plasma electron concentration and its value within anode region

n_α – concentration of particles of α kind in plasma

$\dot{n}_\alpha$ – rate of formation of particles of α kind per unit volume of plasma

n_ω – refractive index of plasma at laser radiation frequency

Nu_1, Nu_2 – Nusselt numbers for internal and external plasma gas flows in tubular cathode

p, p_C – total pressure of plasma and its value at plasma-forming channel wall

p_{ext} – pressure of environment (external) gas

p_α – partial pressure of α-component of plasma

Pr_t – turbulent Prandtl number

q_C – conductive heat flow from plasma to plasma-shaping channel wall

$q_c^a, \langle q_c^a \rangle, q_c^l$ – local arc heat flow into cathode, its averaged value and heat flow introduced into cathode by laser radiation

$\mathbf{q}^D, q_r^D, q_z^D$ – diffusion flow of energy of heavy particles, its radial and axial components

$\mathbf{q}_\alpha^D, q_{\alpha r}^D, q_{\alpha z}^D$ – diffusion flow of energy of α-component of plasma, its radial and axial components

Q – power of laser radiation in plasma

$Q_a, Q_l \equiv Q^0$ – arc power and initial laser beam power

Q_L – laser radiation power absorbed in discharge region of L length

Q_c^a, Q_c^l, Q_c^{tot} – arc heat input, laser heat input and total heat input into cathode

Q_c^j – power released in cathode body by Joule heating

r, φ, z – cylindrical coordinates

r_b – effective radius of laser beam

r_D – Debye radius

r_F – radius of initial laser beam at waist (focal) plane

$r_{s\omega}$ – surface impedance of metal at laser radiation frequency

R – radius of discharge calculation domain

$R_1, - Z_1$ – radius of outlet hole in tubular cathode (radius of cathode nozzle) and outlet hole length

R_2 – outer radius of tubular cathode tip

R_C – radius of plasma-forming channel

R_σ – radius of current-conducting region of plasma

R_ω – reflectivity of metal for non-polarised laser radiation

$S, \langle S \rangle$ – intensity of laser radiation in plasma and its angle-averaged value

S_c – cathode spot area

S^0 – intensity of initial laser beam

T, T_C – temperature of plasma (temperature of heavy particles) and its value at plasma-forming channel wall

T_a, T_c – temperature of plasma within anode and cathode region

T_e, T_{ea} – electron temperature and its value within anode region

T_{ext} – temperature of environment gas

T^c – temperature of cathode

T_α – temperature of α-component of plasma

$\mathbf{u}, v, w, u$ – velocity of plasma, its radial, azimuthal and axial components

u_{ext} – axial velocity of environment gas

$\mathbf{u}_\alpha, v_\alpha, u_\alpha$ – velocity of α-component of plasma, its radial and axial components

U, U_L – total discharge voltage and potential drop for discharge region of L length

U_α – ionisation potentials of atoms and singly charged ions of plasma gas

$\mathbf{v} \equiv \mathbf{v}_\alpha$ – velocity vector of particle of α kind

v_{Te} – electron thermal speed

$V_a, V_c, \langle V_c \rangle$ – anode and cathode potential drops, averaged value of cathode drop

$\mathbf{w}_\alpha$ – diffusion velocity of α-component of plasma

$\mathbf{w}_e^A, w_{en}^A$ – ambipolar diffusion velocity of electrons, its component normal to plasma boundary

$w_{\alpha\beta}$ – dimensionless relative velocity of colliding particles of α and β kinds

W – flow of plasma enthalpy

z^i – positions of conjugate spots for reflected rays of laser beam within cathode hole

Z_1, Z_{II} – boundaries of laser-arc interaction region

Z_α – charge number of α kind particles

α – kind of particle, inclination angle of internal conical surface of tubular cathode

α_e – coefficient of accommodation of electrons

α_{c1}, α_{c2} – heat-transfer coefficients for internal and external surface of tubular cathode

α_{Te} – electron thermal diffusion constant

β – kind of particle, angle of inclination of reflected rays of laser beam within cathode hole relative to horizontal plane

β_T – coefficient of energy exchange between electrons and heavy particles of plasma

γ, γ_L – local discharge parameter and its integral value for discharge region of L length

Γ_ω – absorptivity of metal for non-polarised laser radiation

δ – parameter determing volume (mass) content of the second gas in binary plasma- forming mixture, delta function

δ_{ls} – Kronecker symbol

Δ_f – shift of laser beam focus in plasma

ΔQ^0 – power of initial laser beam falling on tubular cathode surface

ΔU_α – reductions of ionisation potentials

$\Delta \varphi$ – potential jump on space charge layer

$\Delta \omega_\alpha$ – optical shift of continuous spectrum boundaries for atoms and singly charged ions

ε_c – blackness degree of cathode material

ε_i – ionisation energy of plasma

ε^0 – vacuum dielectric permittivity

$\varepsilon_\omega = \varepsilon_\omega' + i\varepsilon_\omega''$ – complex dielectric permittivity of plasma at laser radiation frequency

ζ – refraction broadening coefficient for laser beam in plasma

η – coefficient of dynamic viscosity of plasma

$\eta_a, \eta_l, \eta_{la}$ – efficiencies of arc, laser and combined (laser-arc) action on metal

η_t – coefficient of turbulent viscosity of plasma

η_{la}, η_{la}' – heat (melting) efficiency and total heat efficiency of laser-arc process

η_α – partial coefficient of viscosity for α-component of plasma

$[\eta_{\alpha\alpha}]_1$ – viscosity coefficient for simple gas of α kind particles in the first approximation by Chapmen-Enskog method

ϑ – scattering angle of colliding particles, angle of inclination of initial laser beam ray with respect to beam axis

ϑ_b – angle of focusing of laser beam

θ_α – statistical sums of atoms and ions of plasma gas

$\kappa_\omega, \overline{\kappa}_\omega$ – laser radiation absorption coefficient in plasma, its value averaged over laser beam cross- section

λ – wave length of laser radiation

μ^0 – universal magnetic constant (vacuum permeability)

$\mu_{\alpha\beta}$ – reduced mass of α and β kind particles

ν_e – total frequency of collisions of electrons with heavy particles

v^* – dynamic velocity

ξ – attenuation coefficient of laser beam in plasma

$\hat{\pi}_\alpha$ – tensor of viscous stresses of α-component of plasma

ρ, ρ_C – mass density of plasma, its value at plasma-shaping channel wall

ρ_c – specific electrical resistance of cathode material

ρ_{ext} – mass density of environment gas

σ – specific electrical conductivity of plasma

σ_0 – Stefan-Boltzmann constant

$\sigma_{\alpha\beta}$ – differential cross-section of elastic scattering of α and β kind particles

$[\sigma_e]_1$ – electron electrical conductivity of plasma in the first approximation by Chapmen-Enskog method

τ_C – viscous friction (tangential tension) at plasma-shaping channel wall

$\tau_{\alpha\beta}$ – frequency of elastic collisions of α and β kind particles

φ – potential distribution in space charge layer

$\varphi_c, \overline{\varphi}_c$ – work function of cathode material, its effective value with Schottky's correction

Φ^c – potential distribution in cathode body

$\chi, \chi^{tr}, \chi^{ion}$ – coefficient of thermal conductivity of plasma, its transport and ionisation parts

χ_c – coefficient of thermal conductivity of cathode material

χ_t – coefficient of turbulent thermal conductivity of plasma

χ_α – thermal conductivity coefficient for α-component of plasma

$[\chi_{\alpha\alpha}]_1$ – thermal conductivity coefficient for simple gas of α kind particles in the first approximation by Chapmen-Enskog method

ψ – volumetric radiative losses of plasma

ω – angular frequency of laser radiation

ω_{pe} – electron plasma frequency

ω_α – effective frequencies of levels allowed integratedly for atoms and singly charged ions

$\Omega_{\alpha\beta}^{(l,s)}$ – generalised integrals of Chapmen-Cowling

Introduction

Improvement of the processes of material joining and treatment is a necessary condition for industrial progress. Therefore, the importance of developing new, more efficient methods for welding, cutting, surfacing, etc., based on use of an arc discharge as the least expensive heat energy source, can hardly be overestimated. Unfortunately, further development of arc technologies is hampered by considerable difficulties associated with relatively low concentration of energy in the electric arc plasma and its unstable burning at a high speed of travel with respect to a workpiece. In this connection, the arc discharge used as a technological means fails to meet the increasing requirements imposed on the productivity of metal treatment processes and the product quality.

Another technological means that has been considerably upgraded within a comparatively short period of its utilisation for welding, cutting and various kinds of surface treatment of metals is laser radiation. Thanks to the high concentration of energy in a focused laser beam and the possibility to locally affect a workpiece the laser provides high process productivity and precision. Also laser technologies, unlike, for example, electron-beam ones, possess such advantages as the possibility to carry out a process under atmospheric pressure, as well as to transfer the laser radiation energy over long distances.

However, despite all these advantages, capabilities of the laser as a heat source are limited to a considerable degree. First of all, this is related to the low efficiency of laser radiation in heating metals, caused by their high reflectivity (prior to the formation of a "key-hole") at the wave length bands typical for the majority of industrial lasers. Another factor decreasing the efficiency of utilisation of high-power CO_2-lasers is an over-surface plasma induced by the laser radiation, which greatly reduces the share of the energy introduced by the laser beam into a workpiece. Available are various techniques that can be applied to avoid the above drawbacks, such as, for example, the use of absorbing coatings, and gas jet fed into the treatment zone to blow out the plasma plume. But in such cases, despite some increase in the efficiency of laser methods for metal welding and treatment, the laser is an insufficiently effective and rather expensive device for realisation of the majority of the said technologies.

Hybrid or combined laser-arc processes for material joining and treatment that emerged in the late seventies and were based on the simultaneous application of laser radiation and electric arc presented a new step forward in solving the above problems. In practical realisation of the laser-arc processes both heat sources, as a rule, affect a metal within the common heating zone. Besides, the arc can burn both on the side of a workpiece, to where the laser beam is directed, and on the opposite side. In both cases the electric arc heating of metal leads to an increase of its temperature and, as a result, to an increase in its absorptivity. Thus, the use of the additional arc heating promotes an increase in the efficiency of the corresponding laser process, this being especially important for the application of low-power lasers.

In the case when the arc discharge and the laser beam in the hybrid process act on the same side of a workpiece, the over-surface plasma generated by the laser radiation becomes the root of the arc at the workpiece surface. This leads to an improvement in the spatial stabilisation of the arc root at the metal surface and to a higher stability of the arc burning at low currents and at high speeds of displacement relative to a workpiece. As a result, the application of the electric arc combined with the laser beam allows a considerable improvement in the productivity of the arc welding and heat treatment methods.

When the focused laser beam passes through the arc column, in addition to the above-mentioned phenomena taking place at the surface of a material treated, it causes a change in the entire energy balance of the arc discharge associated with the extra energy evolved in the bulk of the arc plasma due to absorption of the laser radiation. In the case when the power introduced into the arc by the laser beam is commensurate with its electric power, there occurs an intermediate (between optical and arc) type of gas discharge, i.e. the combined laser-arc discharge. The integral and especially local characteristics of the combined discharge plasma will greatly differ from the corresponding characteristics of the electric arc plasma. In its turn, the laser beam interacting with it will also be essentially changed as a result of spatial redistribution of the radiation intensity, as compared to the initial one. Hence, the energy introduced into a workpiece by the laser-arc treatment will not be reduced to a simple sum of the energy inputs of the laser and arc heat sources taken separately. Therefore, the comprehensive study of the combined effect on materials requires an allowance for the mutual influence of the laser beam and the electric arc.

There are a lot of practical advantages of laser-arc welding and allied processes, although not all of the possible hybrid processes are investigated enough until now. The known advantages of the hybrid welding, for example, are the following:

Welding speed. Compared to pure laser welding with filler wire it is possible to significantly increase the welding speed in laser-arc welding. Thus when welding shipbuilding steels of a thickness of 5 mm at a laser output of 12 kW welding speed of 3.2 m/min have been achieved. Reference values obtained by laser welding are about 2.3 m/min and with one-side submerged arc welding about 0.8 m/min.

Capacity for bridging gaps. Although especially in laser welding without filler wire extremely high demands are placed on edge preparation and gap tolerances, plasma cutting of edges does not cause any problem for the hybrid laser-arc procedure. Applying a sensor for detecting the groove geometry and having a corresponding adaptation of the welding parameters, air gaps between 0 and 1 mm can be welded. The arc process increases the ability of gap bridging not only by its filler material but as well by the wider process zone. The arc power determines the width of the weld. The power of the laser determines the welding depth.

Deformations of the component. Despite the absence of systematic investigations, experiments have shown that the deformations occurring during hybrid welding are comparable to those occurring during laser welding.

Weld quality. Whereas the mechanical-technological quality values of the weld often cause problems especially with laser welding without filler metal, the values given in relevant shipbuilding regulations could be achieved with the hybrid procedure without difficulties.

Process stability and efficiency. Experiments with beam paths varying in the welding process between 1 and 20 m showed that the hybrid process is running stable even with moved optics. The laser welding is related to the formation of a "key-hole" due to laser induced plasma. This plasma reduces the ignition resistance of the arc. Therefore the arc itself shows greater stability. Further it is known that the synergistic process leads to an increase of the hybrid welding efficiency.

Furthermore we have to mention that hybrid and combined welding processes can lead to:

– Metallurgical advantages especially in the case of welding stainless steel
– Increase of weldable material thickness
– More weldable materials

– Decrease of filler material consumption
– Decrease of the maximum hardness in the weld metal
– Improvement of weld quality, e.g., crack free weldments
– Possibility to influence the chemical composition of the weld metal
– Improvement of corrosion resistance of weldments
– Decrease of the thermal load on the structure
– Smaller laser power and decreasing investment and running costs

Nevertheless the successful combination of the two processes is no easy matter, depending as it does on a critical set of process parameters. There are a lot of discrepancies in the literature about the advantages and disadvantages of the hybrid processes in connection with the lack of knowledge we have until now. We have to also take into account the strong interaction and correlation between the laser beam and the arc. This leads in some cases to instability of the hybrid processes and to decreasing efficiency. Reasoning from this, further elaboration of the available methods for laser-arc treatment of metals, as well as development of new hybrid processes and technologies are impossible without a deep insight into the mechanisma of laser beam-arc plasma interaction and their combined effect on the metal being treated.

Chapter 1 reviews and classifies the existing methods for combined welding and allied processes, which are based on the principle of a joint use of the electric arc and the laser beam.

Chapter 2 is devoted to theoretical investigations of stationary combined discharge induced by the focused laser radiation of CW CO_2-laser affecting the DC arc plasma.

In Chapter 3 the main principles and methods for calculation of integrated laser-arc plasma torches for various technological processes are described. As an example, the detailed simulation of the integrated plasma torch for laser-plasma powder deposition is included.

Chapter 4 highlights the practical advantages of laser-arc methods for welding and allied processes in comparison with the conventional arc and laser ones, as well as depicting the current state-of-the-art in the field of practical use of hybrid processes.

CHAPTER 1

Development of the Combined Laser-Arc Processes for Joining and Treatment of Materials (Survey)

The number of original papers on laser-arc material joining and treatment processes is continuously increasing and it is necessary to systematise and generalise their results. Therefore, the purpose of the present chapter is to analyse these publications describing both the background of laser-arc methods for welding, cutting, surface heat treatment and coating as well as the state-of-the-art of the issue under consideration.

1.1 Effects of the Simultaneous Action of the Laser Beam and the Electric Arc on a Workpiece

Development of laser-arc processes was started late in the seventies and early in the eighties. The idea was suggested by Steen, an English scientist, to use jointly a laser beam and an electric arc for welding and other kinds of metal treatment, so that both heat sources would affect the metal within the same heating zone. Steen protected his invention by a number of patents [1–4]. He suggested methods for welding, cutting, drilling and surface treatment, wherein the laser beam was directed on to a workpiece and, at the same time, the arc was excited between an electrode and the workpiece. Also, together with other scientists, he carried out the initial experimental studies on the effects of laser-arc action on a metal [5–10] and applied the combined welding method for a particular production process [11]. Practically all specific features of the new laser-arc processes are presented and compared with the cases of the arc and of the laser ones in [5, 6 and 8].

A schematic diagram of an experimental machine for the combined welding and cutting of metals is shown in Figure 1.1. A 2 kW CW CO_2-laser was used in the experiments. With the help of a polycrystalline lens of KCl with a focal distance of 75 mm, a 1 cm diameter laser beam was focused at a workpiece into a spot measuring 250–350 microns. The beam passed through a hole in a copper nozzle of 3 mm diameter in the welding experiments and 1 mm diameter—in the cutting experiments. Helium or oxygen was fed through the laser head nozzle to the work zone for welding or cutting, respectively. A device for TIG welding at currents of up to 250 A (or a welding generator up to a current of 400 A in a set with an oscillator) was used as the arc power supply. The arc torch could be installed either on the same side of the specimen as the laser beam or directed on to the opposite side. A tungsten electrode of the torch shielded by an argon flow served as the cathode in all the experiments, while the workpiece acted as the anode.

The following effects were observed during laser-arc welding and cutting processes. If both heat sources were located on the same side of a workpiece, the electric resistance of the arc burning in the stable mode (current—100 A) decreased under the effect of the laser radiation, this being evidenced by a drop in voltage in the arc gap with a simultaneous increase in current (Figure 1.2a). Besides, the anode region of the arc was "located" on the plasma plume formed over the metal by the laser beam. At lower currents (70 A) the unstable arc burning due to wandering of the anode spot changed to stable (Figure 1.2b) because of stabilisation of the arc anode region within the zone of the thermal action of the laser

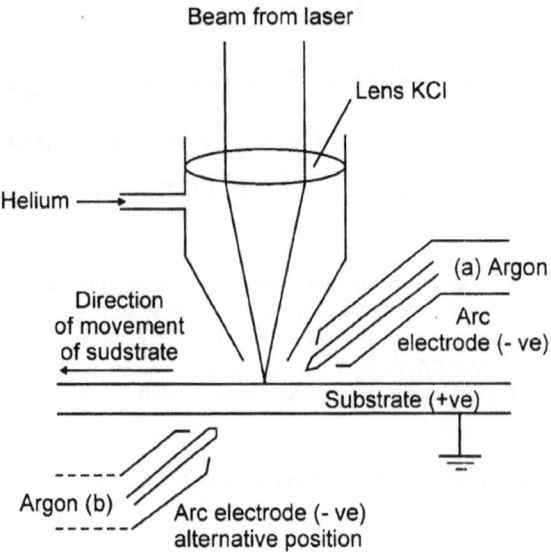

Figure 1.1 Experimental arrangement for laser and arc on the same side of the substrate (a) or on opposite sides (b) [8].

radiation. The latter effect was observed also in the case when the heat sources were located on opposite sides of a workpiece, provided that the temperature of metal under the laser heated spot was in excess of 400 K above the temperature of a surrounding surface.

The effects of the "location" of the anode spot of the arc and the stabilisation of its burning in the combined process even at low currents allowed the welding speed to be significantly increased, as compared not only to arc welding but also to the laser case as well. Figure 1.3 shows the welding speed against the arc current at complete penetration of a specimen (the laser radiation power values are given in the figure). The maximum increase in the speed (4 times) was observed in the welding of low-carbon steel 0.2 mm thick (curve b) with the arc and the laser beam located on the opposite sides of a workpiece. More modest results (an approximately twofold increase in the speed) were obtained for titanium 0.8 mm thick (curves a) and low-carbon steel 2.0 mm thick in welding with the arc and the laser beam located on the same side of a workpiece. Another very important effect was observed in the latter case, namely, the absence of undercuts typical for high-speed arc welding.

An essential increase in the treatment speed, as compared to that in the case of a laser alone, was achieved in the experiments on laser-arc cutting of low-carbon steel 3 mm thick with the laser beam and the arc located on the opposite sides of a workpiece. Figure 1.4 shows dependencies of the cutting speed on the value of the heat input to a metal. The slope of the experimental curve did not change until the heat input to a workpiece due to the arc reached the value that was approximately equal to the heat input due to the laser radiation (1870 W), and cutting took place in such a way as if it was simply because of the increase in the laser radiation power. In this case the quality of cuts persisted and remained comparable with that provided by laser treatment. Above the said limit of the power input (approx. 3800 W), the growth of the cutting speed decreased and the quality of the cut degraded markedly, i.e. the cut was no longer parallel and widened on the arc side because its lateral surfaces were melted by the arc. Both heat sources affecting a workpiece on the

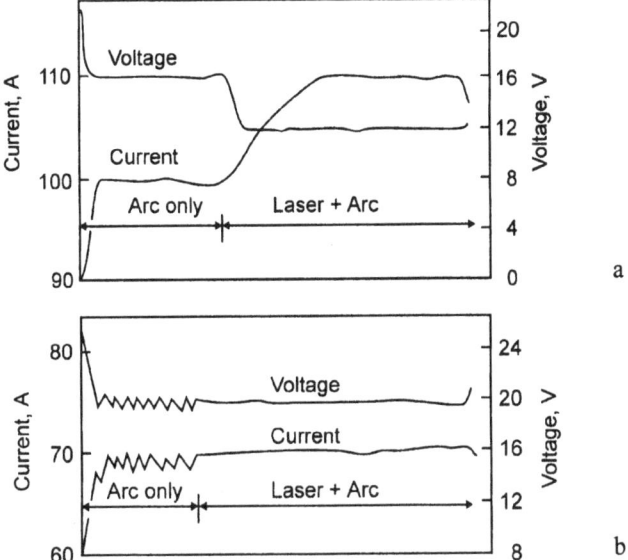

Figure 1.2 Current-voltage characteristics of laser-arc traces: (a) decrease in arc column resistance because of laser, 3 mm thick mild steel at 22.5 mm/s; (b) stabilisation of an unstable arc by laser, 2 mm thick mild steel at 45.0 mm/s [6].

same side also could not provide the cutting quality, because in this case the arc wandered on the cut edges.

The authors of papers [5, 6, 8] made an attempt to interpret the combined effects on metal observed in laser-arc welding and cutting on the basis of physical concepts. In particular, in [5] they assume that the decrease of the arc voltage under the effect of the laser radiation (see Figure 1.2a) was caused by decrease of the anode work function in the zone of the heating spot formed by the laser beam, and they relate this both to the anode potential drop and to the effect of the "location" of the anode region of the arc on the laser heating zone. It is hard to agree with such an opinion, since the anode work function determined by the Fermi energy of electrons in the metal depends only slightly on the temperature of the metal, and on the other hand, is not associated practically with the anode potential drop in the electric arc. In [6, 8] the authors no longer come back to the above assumption and explain the phenomenon of the "location" by the fact that in the one-sided arrangement of the heat sources the arc behaves in accordance with Steenbeck's minimum principle, namely: since electrical conductivity of the laser plasma, whose temperature can reach 20000 K, is much in excess of that of the surrounding cold gas, the laser plasma plume is a preferable, energy favourable region for the path of the burning arc. While agreeing with such an explanation of the effect of the "location" of the arc anode region, it is worth noting that the interaction taking place in this case between the CO_2-laser radiation and the near-anode arc plasma leads to an elevation of its temperature and, hence, the electrical conductivity. It is this circumstance that is the reason for the decrease in voltage in the arc gap mentioned in [6, 8]. It can also be assumed, that with the arc and the laser beam located on different sides of a workpiece, the stabilisation of the anode spot within the laser heated metal zone will not lead to a decrease in the arc voltage (see Figure 1.2b) due to the absence of a direct interaction between the laser beam and the arc plasma.

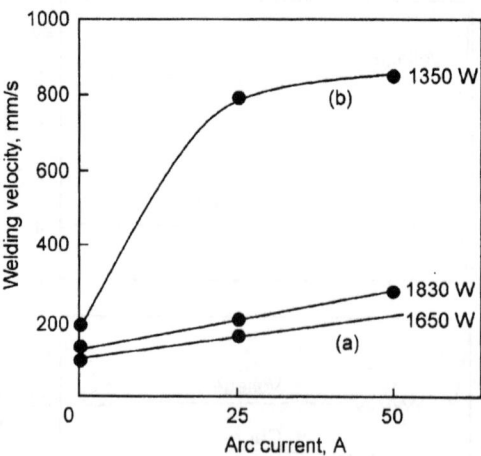

Figure 1.3 Welding speed against arc current: (a) 0.8 mm commercial purity titanium with both laser and arc above sheet; (b) 0.2 mm mild steel with laser above and arc below [6].

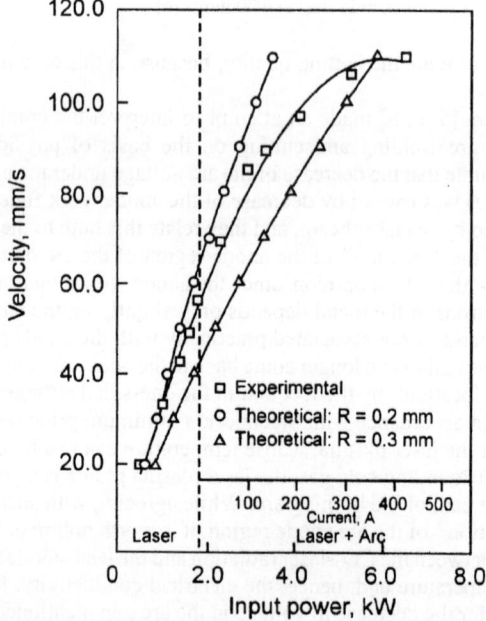

Figure 1.4 Graph of cutting speed against input power. Input power up to 1870 W is by the laser only; above 1870 W the extra power is supplied by the electric arc. Material being cut is 3 mm mild steel [8].

It is stated in [6, 8], that the effect of the "location" of the anode region of the arc discharge on the laser heating spot appears at low and medium arc currents (< 300 A). At high currents the said effect can disappear because, in the authors' opinion, of the self-sta-bilisation of the arc due to a high-power cathode jet, which results in the possible shift of the anode spot independently of the heating zone created by the laser beam. This phenomenon is considered to be one of the reasons for the degradation of the quality of combined cutting with the increase in the arc current [8]. It is supposed also, that the increase in the arc current can result in an undesirable effect appearing even when "location" of the arc anode region on the laser plasma plume occurs (the heat sources are arranged on the same side of a workpiece). This is an effect due to reflection of the laser beam by the plasma, which is observed to occur with an increase in the electron density in the over-surface plasma above the critical value (approx. 10^{19} cm^{-3} for CO_2-laser radiation) and leads to limitation of the heat input to the metal under treatment. The authors of the said papers consider this effect insignificant at arc currents of up to 300 A.

The phenomenon of the "location" of the arc anode region on the laser plasma plume does not only provide an improvement in the arc discharge stability in the combined process, but also leads to such an important effect as the contraction of the arc anode region. The authors of [6, 8] come to this conclusion as a result of comparative analysis of the cross-sectional profiles of the welds (Figures 1.5, 1.6) made in laser-arc welding experiments (dashed curves) and calculated on the basis of a mathematical model of metal penetration (solid curves). The local (within the common heating zone) energy input by the laser and arc heat sources to a workpiece, observed under the arc anode spot "location" conditions, persuaded the authors, when developing the mathematical model, to consider them to be two coaxially acting cylindrical heat sources moving with respect to a specimen. The model suggested that the heat power introduced into a workpiece by the laser beam had a Gaussian distribution, and that introduced by the arc—a parabolic one, and they allowed for the heat losses at the surface of a metal treated due to convection and radiation. It was assumed also, that the effect of "key-hole" penetration occurred in the combined process, like that in laser welding. To allow for this effect, in the mathematical model treating a specimen with penetration, a portion of the calculation domain lying above the isothermal line specifying the boiling point of the metal was considered partially transparent for the laser radiation. It was supposed also, that the power of the laser beam in the "key-hole" cavity decreased with distance from the workpiece surface following the exponential law [8].

By comparing the experimental and calculated profiles of a melt in Figure 1.5 (values of the parameters are given in Figures 1.5 and 1.6) one can ensure that calculation of the shape of penetration into the metal with the arc heat source of radius 0.5 mm and power introduced into a workpiece of 750 kW yields a weld profile as close as possible to that observed in the experiments on laser-arc welding using an arc with a total power of 1000 W. At the same time, referring to the data in the literature, the authors of [6, 8] note that the free-burning arc of such power has effective thermal radius of 1.5 mm. Similarly, the 1600 W arc (with effective thermal radius of 2.0 mm in the absence of laser radiation) in the combined process creates a melt profile corresponding to the calculated one for the heat source radius being 1.0 mm and the effective power—1200 W (see Figure 1.6). By comparing these data the authors concluded that the anode region of the arc is contracted under the effect of the laser radiation, the degree of contraction decreasing with an increase in the arc current.

An explanation of the above effect suggested in [8] is based on the fact that contraction of the anode region of the arc can result from the action of two factors. On the one hand, it is a high-temperature region of the plasma with a higher concentration of the charged particles forming near the arc axis under the effect of the focused laser radiation. On the other hand, it is a jet of metal vapour with a lower ionisation potential than the ambient gas,

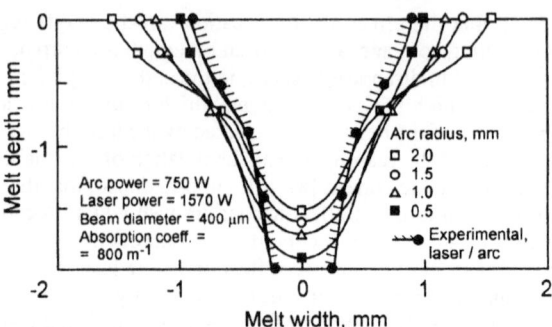

Figure 1.5 Variation in laser/arc melt width and depth as a function of arc radius [6].

emitting from the laser-heated spot. A local increase in electrical conductivity of the plasma under the effect of both factors leads to a concentration of current in the near-axis zone of the anode region of the arc, i.e. to its contraction. In our opinion, when using the CO_2-laser radiation in the combined processes, as in [5, 6, 8], both mechanisms can be considered commensurable contributions to the resulting effect of the contraction. However, the relative significance of each of them can vary with a variation of the laser radiation wave length. Such a conclusion is suggested from the results of experimental studies on laser-arc welding of stainless steel, conducted by one of the authors of this book, using the solid-state Nd:YAG-laser (power—1000 W) and argon-arc non-consumable electrode welding (current—up to 100 A). The main cause of the arc contraction effect observed in this case as well can be the intensive evaporation of metal from the laser action zone, since the formation of a high-temperature plasma region near the arc axis maintained by the laser radiation is highly improbable due to the low coefficient of absorption by the plasma of the laser radiation with 1.06 micron wave length (about two orders of magnitude lower than that of the CO_2-laser radiation).

It is shown in [6, 8] that the weld profiles in the combined process are close to the calculated penetration profiles, if the arc efficiency is selected equal to 0.75 (see Figures 1.5, 1.6), while it amounts only to 0.5 for the conventional arc heat source. It is indicative of the more efficient transfer of energy of the electric arc to metal under the effect of the laser radiation, as a result of which the thermal effect of the combined action is in excess of the simple sum of the thermal effects of the laser and arc heat sources taken separately. Besides, coincidence of the experimental weld profile with the predicted one (Figure 1.6) for the case when the only heat source is the laser beam having a power of 2320 W equal to the sum of powers of the laser source (1570 W) and the arc one at arc efficiency of 0.75 (0.75×1000 W) suggests that the arc, while contracting, acts in the combined process as a focused laser radiation. The effects mentioned above are connected in [8] by the anode region of the arc penetrating the "key-hole" cavity formed by the laser beam. While not rejecting the important role of the phenomenon of the near-anode plasma penetrating the "key-hole" in laser-arc welding and cutting, it should be noted that an increase in the efficiency of the thermal effect of the arc on a workpiece will be observed also in the combined processes in the absence of deep penetration. From our viewpoint, the basic cause of the increase in the arc heating efficiency is a change in a character of the anode processes that takes place under the effect of the laser radiation and, hence, an increase in the heat introduced into the anode by the arc. It is for this reason that this effect is present not only in laser-arc welding and cutting, but also in surface heat treatment as well.

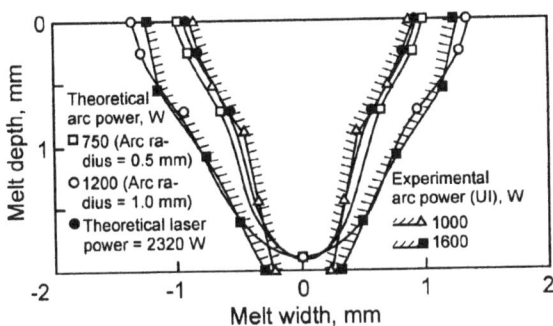

Figure 1.6 Comparison of the predicted and experimental fusion zones associated with 1.0 kW and 1.6 kW of arc augmented power on a laser power of 1.57 kW. Assumed arc power transferred is 75 %, welding speed 33.5 mm/s and absorption coefficient of incident power in "key-hole" plasma 800 m^{-1} [6].

When completing the review of [5, 6, 8], we emphasise an important conclusion the authors came to, based on the analysis of peculiarities of the combined actions on metals in laser-arc welding and cutting. This conclusion implies an interrelation existing between the laser and arc heat sources that leads to a disturbance of the additivity of the thermal action of the laser beam and the electric arc on a workpiece. The latter means that the combined laser-arc heat source can provide a higher rate of metal treatment, than the arc or laser ones taken separately, when each of them have effective power that is equivalent to their total effective power with no interaction taking place between them. In other words, the combined use of the laser beam and the electric arc makes it possible to achieve the same productivity of the process as in the case of conventional laser welding or cutting, but at a much lower power of the laser radiation (naturally, power of the laser radiation should be sufficient to provide stabilisation of the anode region of the arc within the laser heating spot and its contraction). Furthermore, the combined use of the laser beam and the arc discharge allows the metal treatment process productivity to be increased due to a much cheaper energy source of the electric arc, rather than the expensive energy source of laser radiation. This is one of the basic advantages of laser-arc methods for joining and surface treatment of metals.

1.2 Schemes for Practical Realisation of Laser-Arc Welding, Cutting and Surface Treatment

Late in the seventies, following the English scientists, the Japanese researchers started an intensive development of various methods for laser-arc welding, cutting and surface treatment, as well as devices for their realisation. Thus, in 1977, Hamasaki suggested that in TIG welding a focused laser beam should be directed through a special hole made in the arc torch body to a welding pool (somewhat behind the arc in the welding direction) [12]. He patented the method of welding, wherein a combined action on a workpiece was ensured within the common welding pool of the laser beam and the electric arc (laser + GTA).

Early in the eighties, within just two or three years, there appeared a number of unique designs in Japan for practical realisation of the laser-arc processes of metal treatment. Following the chronology of progress in new treatment methods that jointly use the laser beam and the arc discharge, one should note that as early as in 1982 a team of Japanese scientists set up an idea to apply focused laser radiation as an additional heat source for gas-shielded consumable-electrode arc welding [13]. However, the laser beam was used only as an auxiliary tool for postweld heat treatment. For this it was proposed to install a laser head behind an arc head at some distance from it and to weave the former across a weld during welding.

At the same time, another team of Japanese scientists was also inventing the combined laser-arc process for consumable-electrode welding. They suggested that during laser welding a filler wire should be fed to the over-surface plasma plume somewhat ahead of the laser beam [14] and that it should be preheated using an extra heat source (a pilot arc, a gas torch flame, etc.) [15]. Conceptually, the next variant of laser welding using a filler, was provided by preheating the rear side of the weld using the electric arc [16], involving a scheme of the combined laser-arc welding process with the heat sources located on opposite sides of a workpiece, since the laser beam and arc heat-affected zones were joined within the common welding pool.

Further, keeping to the chronology of the emerging combined processes of metal joining and treatment using laser radiation and the electric arc, noteworthy work [17] came out, where the authors suggested placing a filler metal in the gap of a butt joint during laser welding, first melting it with a non-consumable electrode arc torch and, then, carrying out laser remelting by displacing the molten region of the weld beneath the laser head without waiting for the pool to cool down. Apparently, as well as in [13], the realisation of the combined laser-arc process has not yet occurred, since here a simultaneous action of the laser and arc heat sources on a workpiece is lacking. Note that [18] also deals with the simultaneous use of the arc with a non-consumable electrode and laser radiation for the combined welding process. Actually, it is suggested in this work to apply the laser-arc welding scheme known from the papers by Steen (see, e.g. [6]), where the laser beam and the arc were located, respectively, on the face and back sides of a workpiece to make circumferential welds in pipelines with a view to eliminate excessive weld reinforcement and lack of penetration.

Eventually, Hamasaki, an author of the above-mentioned invention [12], at the end of 1982 offered a welding method [19–21], wherein a metal workpiece was simultaneously penetrated by the arc burning from a consumable electrode in a shielding gas as well as by laser radiation, the laser beam being focused near the bottom of a crater formed by the arc. With this method the laser head and the torch for consumable electrode welding were located over a metal surface in such a way that the laser beam was somewhat ahead of the arc in the welding direction. This invention initiated the development of laser-arc welding using a consumable electrode in shielding gases (laser + GMA).

Approximately a year later an ingenious device was proposed in [22] for realisation of combined non-consumable electrode laser-arc welding (laser + GTA). In this device the laser beam is directed to a workpiece along the axis of the arc torch nozzle, through which a working inert gas is fed (Figure 1.7). Four pairs of tungsten rod electrodes are arranged on a cylindrical insulator coaxially with the beam. A shielding gas is also fed into the nozzle. Welding is performed with the two arcs burning simultaneously from diametrically located electrodes (ahead of and after the beam). Another design provided for using a special internal nozzle serving as an arc cathode (Figure 1.8) instead of a set of rod electrodes (see Figure 1.7).

At the very beginning of 1984 the author of [23] suggested a device for cutting, welding, hardening and spraying that simultaneously used laser radiation and a plasma jet (Figure 1.9). The plasma jet is formed by the arc ignited in an annular gap between two electrodes, the laser head nozzle and the external water-cooled nozzle being located coaxial to it serving as the electrodes. A plasma-forming gas is fed both to the laser head and to the annular gap between the electrodes (nozzles). At the outlet it forms a plasma jet which is coaxial with the laser beam. The device described can be considered to be one of the first indirect action plasma torches designed for realisation of combined laser-plasma jet processes of metal treatment (laser + PJ).

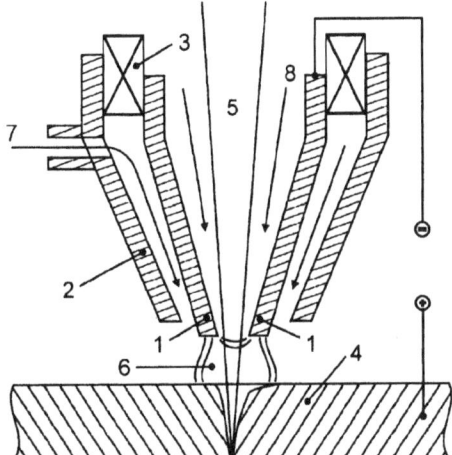

Figure 1.7 Laser-arc welding device: 1, rod electrodes (cathodes) for TIG welding; 2, nozzle; 3, cylindrical insulator; 4, workpiece; 5, focused laser beam; 6, electric arc; 7, shielding inert gas; 8, inert gas [22].

In papers [24, 25] published in the middle eighties the use is suggested of a ring-like electrode fixed at the focusing laser head nozzle end as an arc electrode in a device for laser-arc welding, cutting and surface heat treatment. The authors of this invention believe that this device will be helpful for increasing the efficiency of laser heating of metal, since the arc plasma, while absorbing the laser radiation energy, transfers it to the workpiece in a

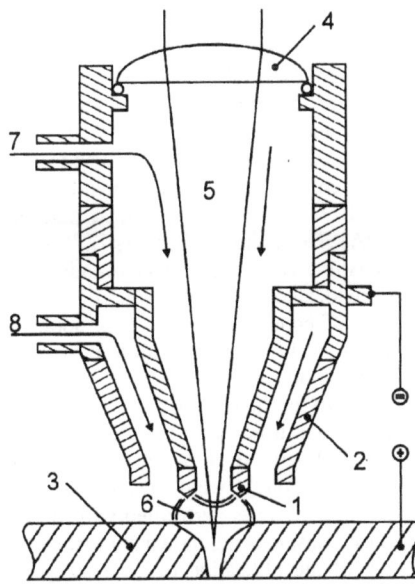

Figure 1.8 Laser-arc welding device: 1, annular electrode (cathode); 2, nozzle; 3, workpiece; 4, lens; 5, focused laser beam; 6, electric arc; 7, inert gas; 8, shielding inert gas [22].

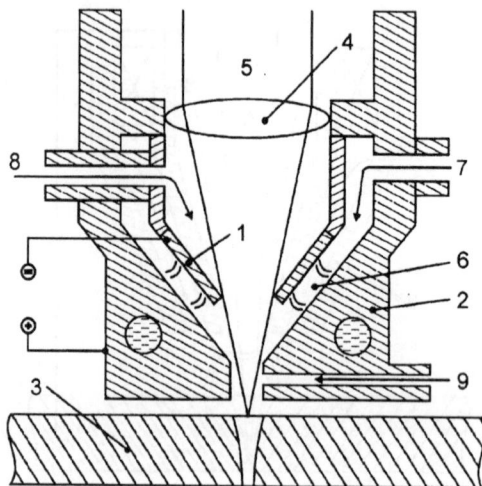

Figure 1.9 Working head in working device using laser light: 1, laser head nozzle (cathode): 2, external water-cooled nozzle (anode); 3, workpiece; 4, lens; 5, laser beam; 6, electric arc; 7, 8, plasma gas; 9, spray particles (for spraying process) [23].

more efficient manner than the laser beam alone. In the authors' opinion, this makes it possible to weld such electrical- and thermal-conducting metals as aluminium and copper. Note that the device suggested, as well as the devices described in [22] allow a special type of the discharge to be realised, i.e. the combined laser-arc discharge, the characteristic feature of which is substantial interaction between focused laser radiation and the electric arc plasma column.

To increase the service life of a ring electrode and reduce the heat-affected zone in laser-arc treatment, the same authors offered an improvement in the device described above provided by the use of an additional external water-cooled nozzle, through which a shielding gas is fed [26]. In addition to protection of weld metal from oxidation by atmospheric oxygen, this gas help cool the electrode and, thus, increases its service life, as well as constricting the arc column. As a result, the heat-affected zone is reduced so, that it makes it possible to use this device to cut metals. Eventually, this improvement led the authors to the creation of the direct action plasma torch, based on the combined discharge, for laser-plasma arc treatment of metals (laser + PA). The author of the work [27] suggested combining the laser radiation and the plasma (transferred) arc to make a two-sided weld by a method, in which the laser beam and the arc were located on opposite sides of the workpiece (on the weld face and back side, respectively). Of cource we may speak about interaction between laser radiation and arc plasma, and also about the formation of the combined discharge, only if through penetration of a specimen occurs.

Another method is also known for the laser-arc process with the heat sources located on opposite sides of a workpiece. It was used as the basis for the method for combined laser + GMA welding of butt joints with V-groove preparation [28]. In this method the arc torch is placed on one side of the groove, i.e. on the weld face, and the laser head on the rear side of the weld. Welding using the laser beam is performed with full penetration, resulting in a considerable amount of the laser plasma being blown out to the groove. Firstly, this provides an increase in the efficiency of the utilisation of the energy of the laser radiation and, secondly, improves the arc burning stability due to the laser plasma present in the groove. The method provides one-pass welding of thick-plate metals using a low-power laser.

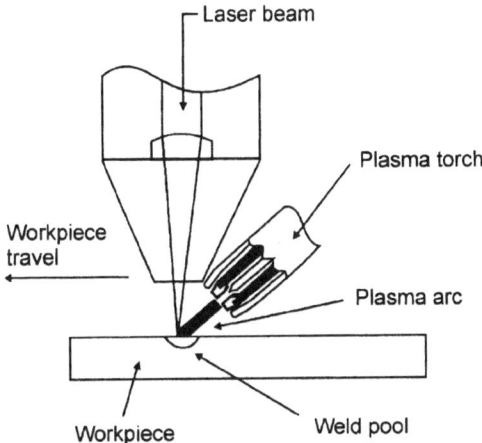

Figure 1.10 Experimental arrangement of plasma arc augmented laser welding system (PALW) [30].

The work on developing new methods for application of laser-arc processes has continued during the last decade. In particular, in the scientific literature of this period we can find information on the method of laser + PJ cutting where heat sources are located on opposite sides of a workpiece [29], and on the method of laser + PA welding which has been realised by following a new approach [30, 31]. Unlike the scheme suggested earlier [26], according to which the laser beam and the plasma arc are in axial alignment with respect to each other, in [30] the axis of the laser beam and that of the plasma torch are arranged at an angle to each other (Figure 1.10). Another practicable scheme for realisation of the process of laser + PA welding providing for feeding a filler wire to the welding zone is described in [32] (Figure 1.11).

When completing consideration of the known methods for practical realisation of laser-arc processes, an earlier review paper [33] is worth noting. Along with the methods

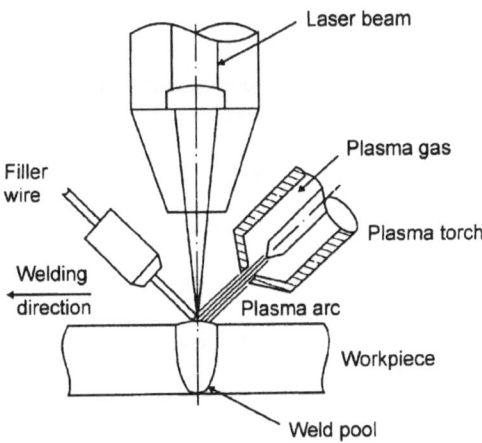

Figure 1.11 Laser + PA welding with a filler wire [32].

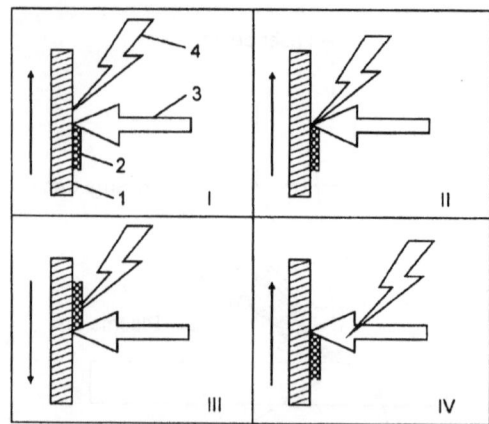

Figure 1.12 Laser-plasma spraying schematic: I, substrate; II, coated powder; III, plasma jet; IV, laser beam. Arrows point to the direction of movement of substrate [33].

for combined laser-arc welding described above, it considers other hybrid processes of material treatment, in particular, plasma-laser spraying. It is emphasised in the paper that the application of laser radiation for glazing a layer deposited with the help of the plasma jet improves the structure, mechanical and functional properties of the resulting coating. Four alternative schemes for the realisation of the combined spraying process (Figure 1.12) are presented in [33]. These methods provide different conditions for interactions taking place between the laser beam and the plasma jet with a sprayed material. The paper indicates that plasma-laser spraying achieves higher productivity than PVD and CVD processes. In our opinion, the latter of the schemes shown in Figure 1.12 is the most promising, because

Table 1.1 Basic schemes for practical realisation of laser-arc processes.

No	Type of combined process	Location of laser beam and arc relative to workpiece	Process application field	References
1	Laser + GTA	On one side of a workpiece	Welding, cutting, drilling, surface treatment	[1, 8, 12, 22, 24]
2	Laser + GTA	On opposite sides of a workpiece	Welding, cutting	[6, 8, 18]
3	Laser + GMA	On one side of a workpiece	Welding	[19, 33]
4	Laser + GMA	On opposite sides of a workpiece	Welding	[28]
5	Laser + PA	On one side of a workpiece	Welding, cutting	[26, 30 – 32]
6	Laser + PA	On opposite sides of a workpiece	Welding	[27]
7	Laser + PJ	On one side of a workpiece	Welding, cutting, hardening, spraying	[23, 33]
8	Laser + PJ	On opposite sides of a workpiece	Cutting	[29]

it allows the laser radiation to be utilised not only to heat a material being sprayed, but also to control the plasma jet characteristics due to the plasma-laser interaction.

The above methods (given in chronological order) of application of laser-arc processes for joining and treatment of materials are classified by the type of combined process and by the spatial arrangement of heat sources relative to the workpiece and are summarised in Table 1.1. It should be noted in conclusion, that the known methods for laser-arc welding, cutting and surface treatment (laser + GTA, laser + GMA, laser + PA and laser + PJ), as well as the majority of the devices for their practical realisation were suggested before the middle eighties. After that, judged by the papers of the last ten years, scientists concentrated their efforts to a greater degree on investigating the technological potential of laser-arc metal treatment methods, as well as on studying the physical processes occurring during the interaction of the laser radiation with the arc plasma and their combined effect on metals.

1.3 Energy Characteristics and Technological Potentialities of the Combined Laser-Arc Heat Source

In parallel with the development of various laser-arc processes and devices for their realisation, an extensive experimental data have been accumulated on the energy characteristics and the technological potentialities of the new combined heat source. The parameters of metal penetration, the productivity and the quality of laser-arc treatment have been investigated, including the ratio of the powers of the laser beam and the electric arc, their relative locations, conditions for focusing the laser radiation, etc. These investigations were aimed at the evaluation of the prospects for industrial application of various combined processes and finding the ranges of their optimal conditions.

As noted above, Steen was the first to study the technological potentialities of his invention. He investigated the efficiency of the simultaneous use of up to 2 kW CO_2-laser radiation and a non-consumable electrode arc for the welding and cutting of low-carbon steels of small thickness (0.2–3.0 mm), for welding titanium 0.8 mm thick, as well as for cladding thick-plate (12 mm) low-carbon steel [5–11]. He found that the combined effect of these heat sources on a workpiece allowed the penetration depth and the process productivity to be greatly increased, practically without any degradation of the quality of treatment, as compared to the use of the laser alone.

The efficiency of the combined heating of a metal being treated by low-power CO_2-laser radiation and the tungsten-electrode argon arc (with both heat sources located on the same side of a workpiece) was also considered in [34, 35]. Experiments conducted on the combined welding of carbon and stainless steels, nickel alloys, brass and copper showed a substantial increase in the penetration depth, as compared to laser welding, in the case of coincidence of the arc crater and laser beam focus positions. For example, in laser-arc welding of austenitic steel pipes 150 mm diameter with 4 mm wall thickness through penetration was achieved with 450 W laser power and 25 A arc current (approx. 190 W) [34]. Without the arc, a laser beam of the same power penetrated metal only to a few percent of its thickness. To provide through penetration the laser radiation power had to be increased to 1 kW. A three and higher fold increase in the penetration depth was observed also in the combined welding of 4 mm thick stainless steel sheets, being evidenced by the dependencies, given in [35], of the geometrical parameters of the weld on the power of the laser beam at variable arc current and constant welding speed. The authors revealed that the effect of additional arc heating on the metal penetration depth decreased with growth of the laser radiation power. With laser power above 400 W a change in the arc heat source power from 180 to 600 W hardly influenced the penetration depth, but led only to widening of the weld.

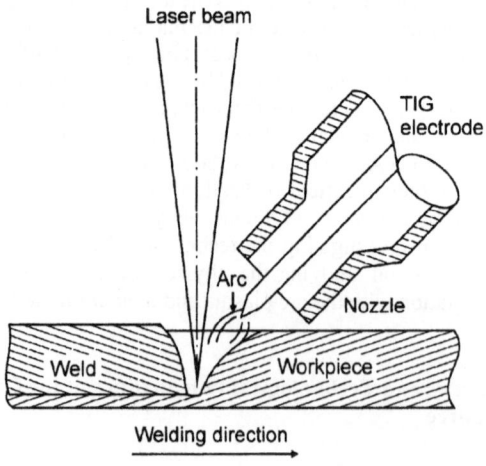

Figure 1.13 Experimental arrangement for laser TIG welding [39].

Moreover, if an arc with a power of above 600 W was used in the combined process, the weld was widened so that it became almost indistinguishable from a weld made without the laser beam. Therefore, it can be supposed that, for each fixed value of the laser radiation power, there exists a limit of the electric arc power, beyond which the weld shape and, hence, the character of the thermal effect on the metal treated are determined mostly by the arc energy source, and the role of the laser beam as one of the components of the combined heat source becomes insignificant. In other words, there exists an optimum ratio between the powers of the laser beam and the arc joined together in a combined process. For example, it was found in [35], that in the case of laser-arc welding of nickel alloy pipes with 2.6 mm wall thickness the best results were obtained with 120 W arc power, the radiation power being 400 W. Such a ratio between the powers allowed the welding speed to be doubled, as well as the weld shape typical for laser welding to be preserved. This ratio is likely to have an optimum for each given process.

Papers [36–40] by Hamasaki and other Japanese scientists deal with the investigation of technological peculiarities and advantages of the combined welding of thick-plate (up to 15 mm) low-carbon steels. The results of these investigations are presented most comprehensively in [39], in a paper of a fairly general character. We will therefore give it special consideration.

Two methods of laser-arc welding were studied, i.e. laser TIG and laser MIG*. Firstly a process scheme was used with a tungsten electrode — cathode (diameter—2.4 mm, angle of inclination to the workpiece surface—45 degrees, interelectrode gap—1.0 mm) located ahead of the laser beam in the welding direction (Figure 1.13). Here the arc melted the upper metal layer and the laser beam provided the "knife" for penetration. In the experiments on laser MIG welding an arc torch with a consumable electrode — anode (diameter of welding low-carbon steel wire—1.2 mm, maximum angle of inclination to a workpiece surface—75 degrees) was placed behind the beam (Figure 1.14). Welding was performed on plates having different groove preparations. In this case, unlike the laser TIG process, the laser

---

* In classification of the combined processes suggested in Section 1.2 they can be referred to as laser + GTA and laser + GMA processes, respectively.

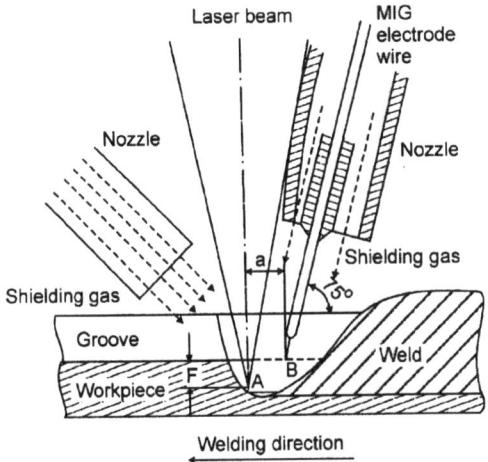

Figure 1.14 Experimental arrangement for laser MIG welding [39].

radiation melted the contacting edges of the joint being welded, while the consumable-electrode arc was used to fill up of the groove.

The results of the experiments on the determination of the penetration depth in laser TIG welding are illustrated in Figures 1.15–1.22. In particular, Figure 1.15 shows the plots of penetration depth against welding speed at variable powers of the laser radiation and constant arc current (300 A). By comparing this figure with Figure 1.16, which shows similar dependencies for purely laser welding, one can confirm that the penetration depth in the combined process significantly increases. For example, with a 5 kW power of laser radiation, it is 1.3– 2.0 times (depending on speed) higher than that provided by laser welding with the same power of the beam. According to the estimates made by the authors of [39], the

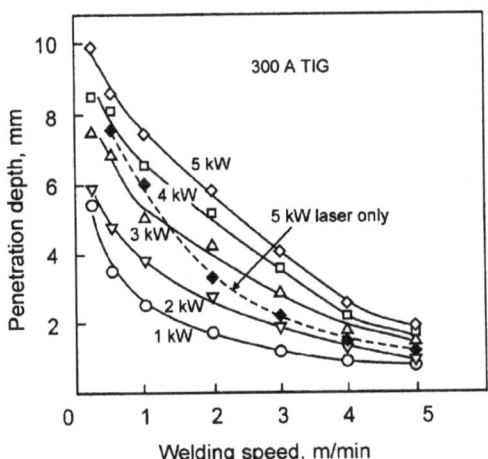

Figure 1.15 Effect of welding speed on penetration depth for laser TIG welding at constant 300 A arc current [39].

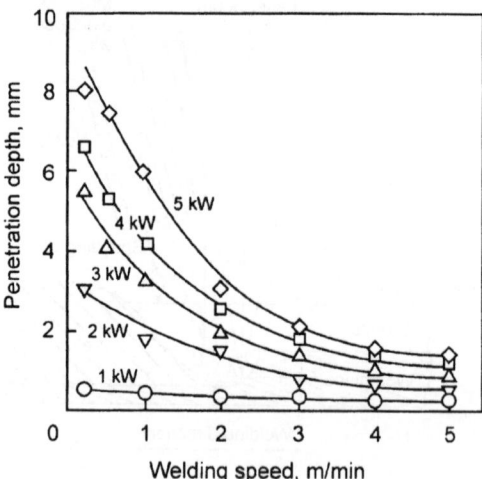

Figure 1.16 Relationship between penetration depth and welding speed for laser welding alone [39].

combined effect of the high-power (5 kW) laser radiation and the tungsten-electrode arc, is higher than in the experiments conducted by Steen [5, 6, 8, 9], who used a laser of a lower power (2 kW). However, this conclusion is disputable, since, as can be seen from comparing Figures 1.15, 1.16, contrary to the above the ratio of the values of the penetration depths in laser-arc and laser welding (which, in fact, is an index characterising the efficiency of the combined process) decreases with increase in the laser radiation power.

Dependencies of the penetration depth on the welding speed were measured for variable arc currents and constant power of the laser radiation (Figure 1.17). Curves illustrating the dependence of the penetration depth on the arc current (Figure 1.18a) and on the laser radiation power (Figure 1.18b) at a welding speed of 1.0 m/min were plotted on the basis of

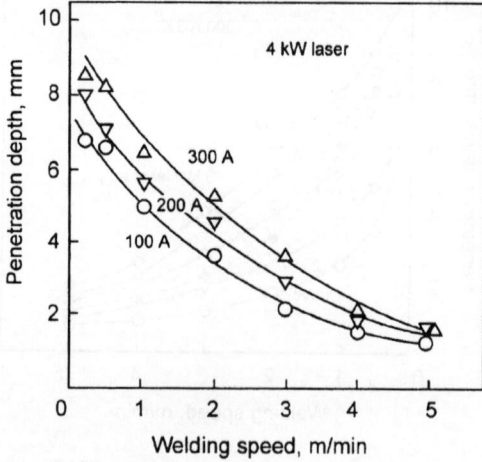

Figure 1.17 Effect of welding speed and TIG arc current on penetration depth for laser TIG welding at constant 4 kW laser power [39].

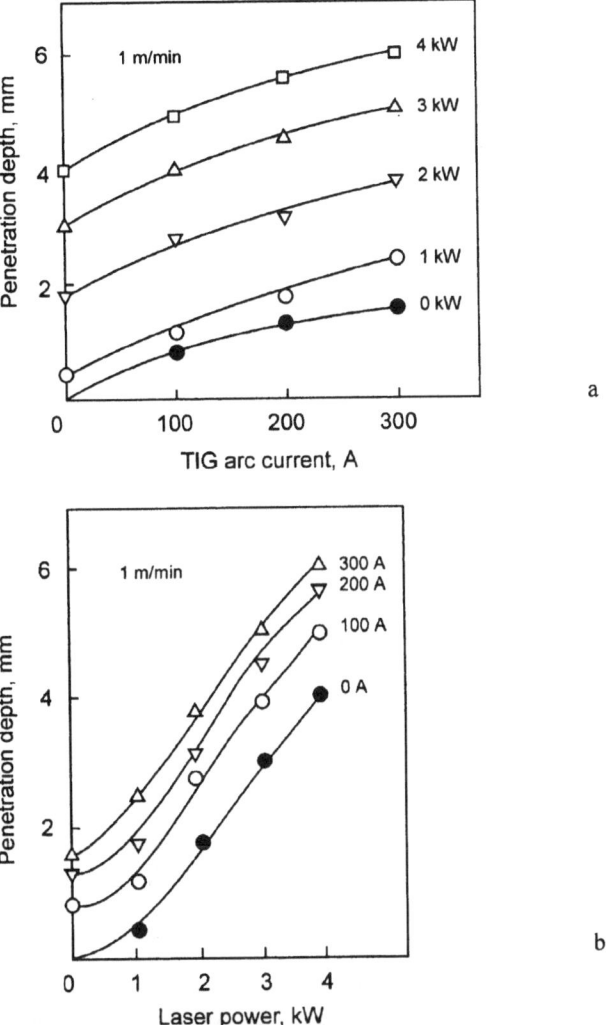

Figure 1.18 Effect of TIG arc current (a) and laser power (b) on penetration depth [39].

experimental data given in Figures 1.15–1.17. These figures clearly prove that the use of the additional arc heating provides the same penetration depth as in case of laser welding, but at a much lower power of laser radiation.

Of interest are the investigations conducted by the authors for determining the optimum distance between the non-consumable electrode and the laser beam axis (Figure 1.19) and the optimum interval of position of the beam focal plane relative to the specimen surface (Figures 1.20–1.22) in laser TIG welding. According to the data in [39], to provide the maximum penetration the distance from the non-consumable electrode tip to the laser beam axis should be in the range of 2–3 mm. For smaller distances there exists a risk of failure of the electrode under the effect of the laser radiation, while with larger distances a decrease in

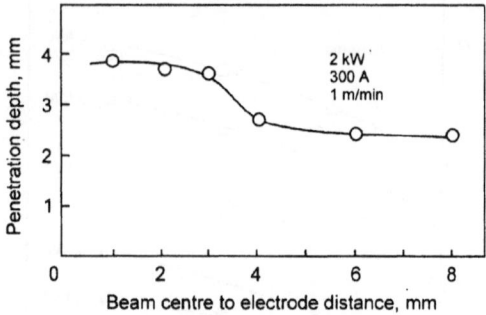

Figure 1.19 Effect of beam center to electrode distance on penetration depth [39].

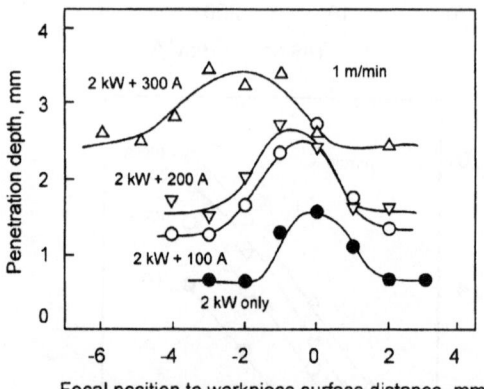

Figure 1.20 Relationship between focal position to work surface distance and penetration depth at constant welding speed [39].

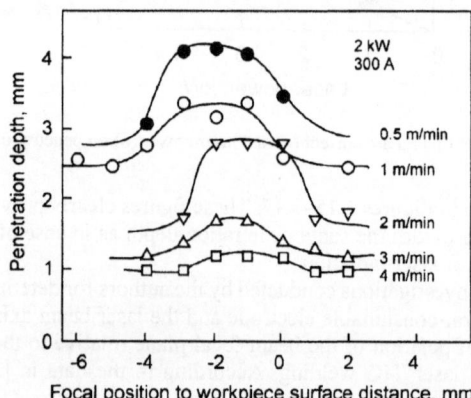

Figure 1.21 Relationship between focal position to work surface distance and penetration depth at constant laser power and welding current [39].

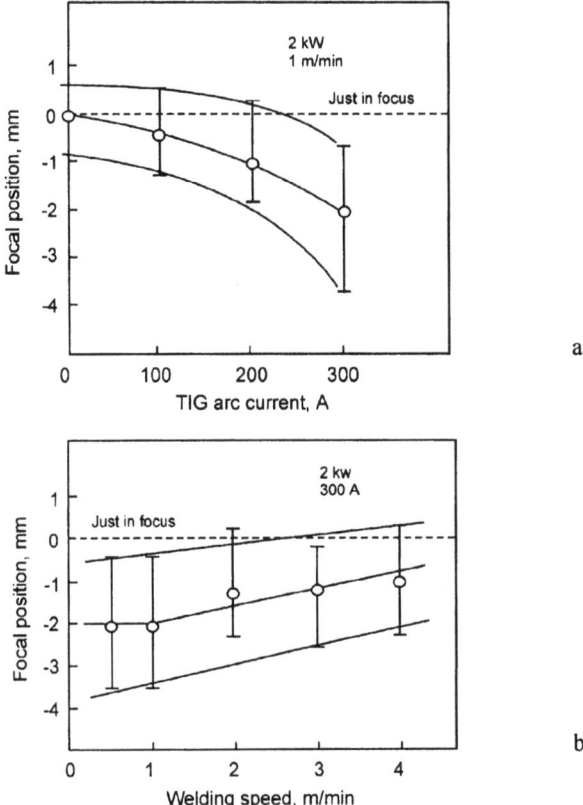

Figure 1.22 Effect of: welding current (a); speed (b) on focal position for maximum penetration [39].

the penetration depth occurs (Figure 1.19), which is caused, apparently, by the disappearance of the effect of the " location" of the arc anode region on the laser heating spot. The optimum interval for the position of the beam focus relative to a workpiece surface (ensuring a penetration depth of not less than 90 % of its maximum value) can be selected from Figure 1.22. It follows from this figure, that the optimum position of the focal plane shifts down from the specimen surface (Figure 1.22a) with increase in the arc current and, conversely, approaches it with increase in the welding speed (Figure 1.22b). This effect is clear, noting that the laser beam focus should track the position of the weld pool surface, which varies according to the growth of the arc current and the welding speed.

Turning to the description of the investigation results obtained by the authors of [39] in laser MIG welding, it should be noted that the aim of the simultaneous use of the laser and the consumable-electrode arc was to check the feasibility of joining 12 mm thick low-carbon steel plates by this method in one pass at a speed of 0.5 m/min. This cannot be done either by consumable-electrode welding at currents of up to 400 A, or by laser TIG welding at a radiation power of 5 kW and an arc current of 300 A. Laser MIG welding was carried out in a helium atmosphere at laser radiation power of 5 kW, an arc current of 400 A and various groove shapes.

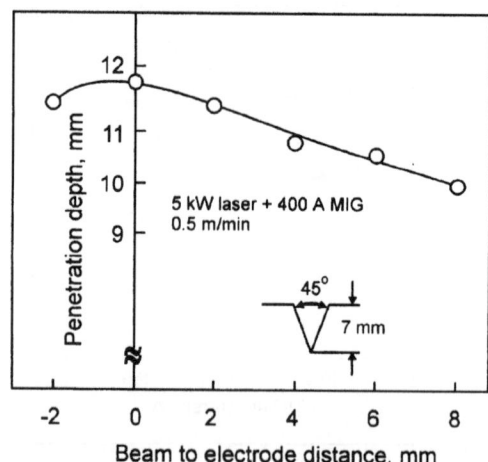

Figure 1.23 Relationship between beam to electrode distance and penetration depth [39].

Figures 1.23, 1.24 illustrate the dependence of the penetration depth, respectively, on the position of the consumable electrode relative to the laser beam (parameter *a* in Figure 1.14) and the position of the beam focus relative to the groove bottom (parameter *F* in Figure 1.14) for different shapes of the latter. It follows from Figure 1.23 that the maximum penetration depth is provided by the intersection of the electrode axis and the laser beam axis at the bottom level of the groove, i.e. at *a* = 0. The optimum focus position is below the said level (see Figure 1.24), this being associated with a sag of the weld pool surface as a result of the forces applied by the arc. Finally, the dependence shown in Figure 1.25 of the penetration depth on the laser radiation power in laser MIG welding, allows the conclusion to be drawn that a 4 kW laser combined with an arc burning in a helium atmosphere at 400 A current can weld 12 mm thick low-carbon steel plates at a speed of 0.5 m/min by using a V-groove to a depth of 7 mm with groove angle of 45 degrees. The authors of [39] note that by increasing the radiation power up to 5 kW thicker plates can be welded at the same speed or 12 mm thick plates can be welded at speeds increased up to 0.8 m/min.

In addition to the experimental data described above, the paper under consideration includes photographs of the transverse microsections of welds made by laser TIG welding at various speeds, arc currents and laser radiation powers, as well as by laser MIG welding, as compared to laser and MIG welding methods. As to the quality of welded joints, both laser TIG and laser MIG welding are reported to provide good weld formation, absence of undercuts and non-uniform reinforcements typical of welds made, for example, by TIG welding under similar conditions, together with the absence of pores and gas holes. This resulted in superior mechanical properties of the joints produced.

It should be noted, when completing the review of papers [36–40], that the authors ascribe a high estimate to the technological potential of laser-arc welding methods. They conclude that these methods are promising for industrial application, because the combined technologies do not require high-power laser units and, at the same time, make it possible to realise the advantages of laser welding with respect to the shape and the quality of welds, as well as from the point of view of process productivity.

Paper [41] published by American scientists in 1984 is, to our knowledge, the first work dedicated to investigation of the technological peculiarities of laser-arc welding of aluminium and its alloys. The authors set themselves the task of interpreting results obtained

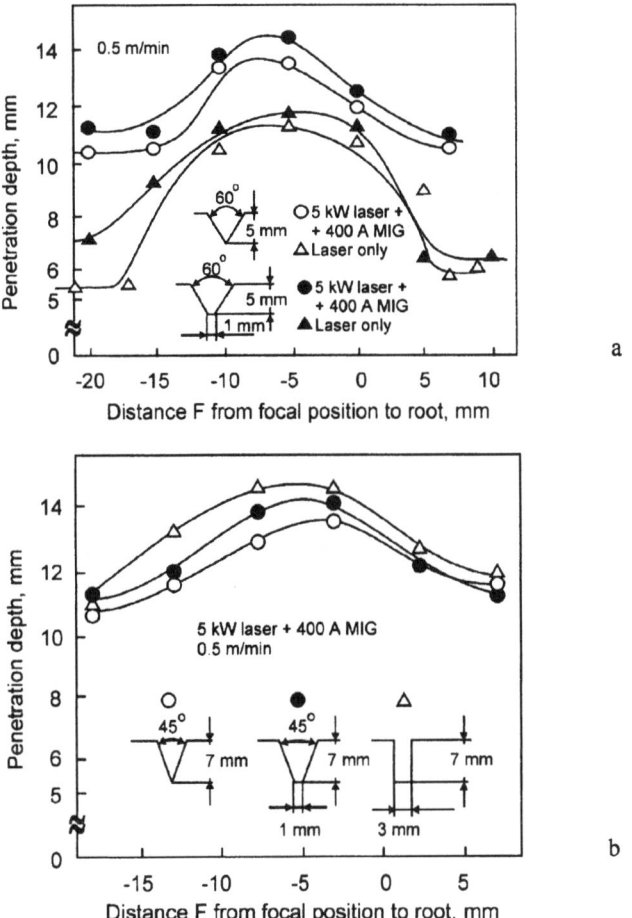

Figure 1.24 (a) and (b) Effect of groove shape on penetration depth [39].

on the basis of physical concepts of the mechanism of the laser beam and arc interaction, in addition to studying the effect for various parameters of the combined process on the penetration depth and the arc burning stability in welding aluminium alloys.

The experiments involved a 600 W CW CO_2-laser and the arc torch with a tungsten electrode (cathode) 2.4 mm diameter installed ahead of the laser beam in the welding direction, so that the angle between the workpiece surface and the electrode was 45 degrees and the distance from the electrode tip to the laser beam axis did not exceed 3.2 mm. Laser + GTA welding was performed with incomplete penetration of butt joints in the 5052 aluminium alloy plates of 3.2 mm thickness. The results obtained were not compared with the results of laser welding, as was the case for the majority of the papers considered, but with those of arc welding of similar specimens, since laser welding could not provide the required penetration due to insufficient power of the laser.

The experimental data given in Figures 1.26–1.28 show an increase in the penetration depth and an improvement in the arc stability in the combined process, as compared to

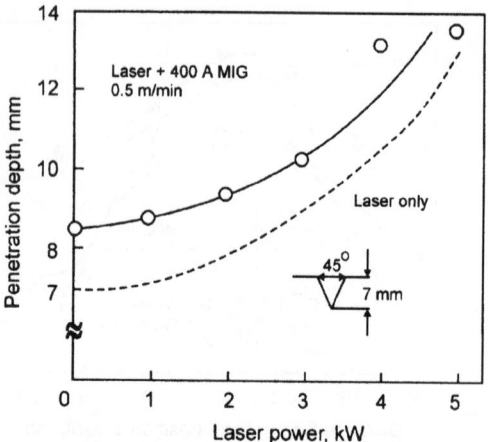

Figure 1.25 Relationship between laser power and penetration depth with and without 400 A MIG arc [39].

tungsten electrode arc welding. In particular, the relative increase in the penetration depth was 20–50 % within the range of the welding conditions applied (compare Figures 1.26 and 1.27). It is the opinion of the authors of [41], this is less than the extra effect of the laser in welding carbon steels and can be explained by the difficulties commonly encountered in welding aluminium alloys by low-power laser radiation. It should be noted, however, that if the data of [39] are taken into consideration (see Figure 1.18a), the effects of the increase in the penetration depth for steel and aluminium can be estimated as commensurate, the other conditions being equal. It is noted also in [41], that despite the increase in the penetration depth provided by the

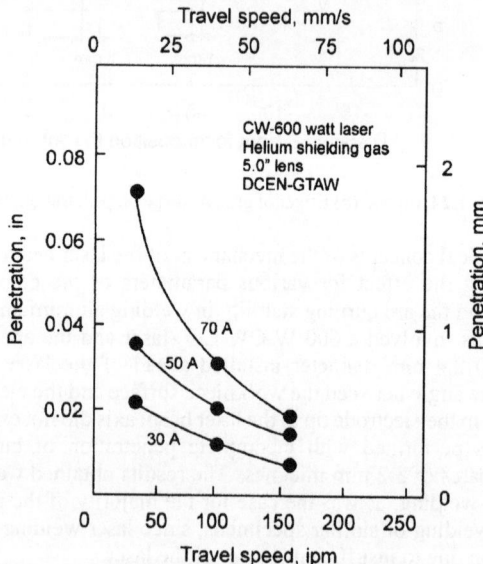

Figure 1.26 Laser + GTA joint penetration vs. travel speed at three current levels [41].

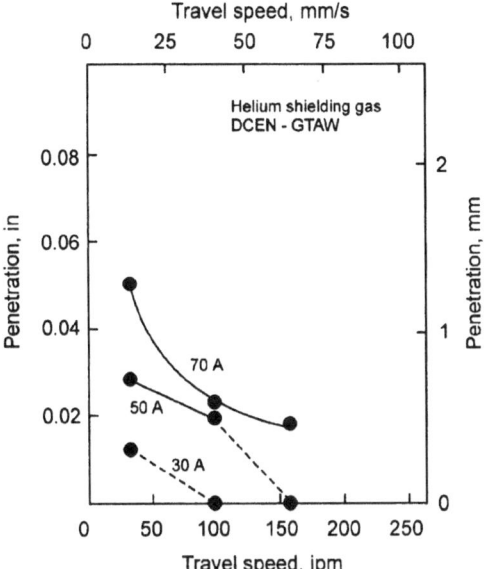

Figure 1.27 GTAW joint penetration vs. travel speed at three current levels. Note: All welds showing no joint penetration were unstable [41].

combined process, the weld shape still remains typical for arc welding at all the arc currents used. It suggests that in laser-arc welding of aluminium alloys the effect of the arc contraction is less evident than, for example, in the combined welding of steels [6, 8]. In our opinion, the reason for such difference in the arc behaviour lies in an increase in the transverse sizes of the laser evaporation zone of metal, determining a degree of contraction of the arc anode region, with an increase in thermal conductivity of this metal.

It was found, when studying the stability of the process of laser-arc welding of an aluminium alloy in comparison with non-consumable-electrode welding, that the presence of the laser radiation causes stabilisation of the arc, and that this leads to the possibility of

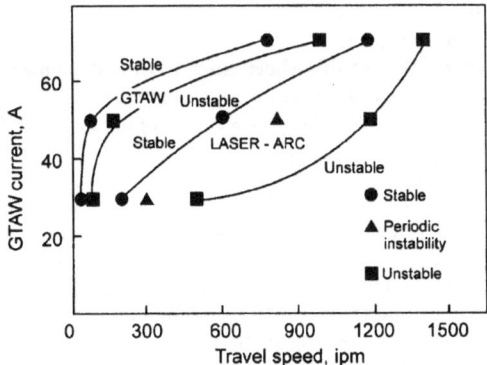

Figure 1.28 Arc current vs. travel speed. CW 600 W laser power, helium shielding gas, 5.0 in. laser lens, DCEN GTAW [41].

an increase in the welding speed, the effect being most noticeable at low currents (see Figure 1.28). The authors managed to produce smooth homogeneous welds at welding speeds of up to 30.5 m/min (1200 ipm) and an arc current of 70 A (see Figure 1.28) with the help of the combined process, whereas arc welding at this current restricted the speed to not more than 20.3 m/min.

Work in reference [41] suggests three possible mechanisms to interpret the observed phenomena. They are: stabilisation of the anode spot of the arc in the presence of the laser beam, an increase in absorption of the beam energy by the workpiece due to the arc heating of the metal under treatment and, finally, direct interaction of the laser radiation with the arc plasma. The authors of [41] think that at the laser beam power of 600 W the last two mechanisms contribute only slightly to an improvement in the stability of the process and an increase in the penetration depth, while the decisive role is played by stabilisation of the anode spot of the arc within the laser heating zone. Generally speaking, one can agree with this statement, but it should be noted that the physical mechanism underlying the stabilisation effect observed by the authors (as well as the phenomenon of arc anode region contraction) is a jet of aluminium vapour with a lower ionisation potential than helium, appearing in the over-surface plasma as a result of the laser evaporation of metal. This can also be used to explain the effect of the arc being more easily excited in the presence of the laser beam, as noted in [41].

Data on the realisation of laser + GTA welding of thin sections of aluminium and its alloys, as well as other thin-sheet materials can be found in later publications [42–45]. The authors of these papers published during the last decade note the so-called synergistic effect of the combined use of laser radiation and the electric arc (the effect of disturbance of additivity of heat of the laser beam and the arc plasma influencing a workpiece, which has been mentioned in Section 1.1). For example, it is noted in [42] that in the experiments on laser-arc welding of aluminium using a 1.2 kW CO_2-laser and a TIG arc at a current of 60 A and welding speed of 25 cm/min, the volume of molten metal was larger by 75 % than the sum of volumes of the melts produced by each of the heat sources used separately. It is emphasised in the above-mentioned references that sound welds can be produced by the combined method provided that the laser beam and the arc are of a relatively low power and that the welding speed is sufficiently high, which was unattainable with laser welding using laser radiation of a much higher power. This advantage of the process was utilised by the authors of the study described in [44], who employed a solid-state Nd:YAG-laser with a power of only 360 W in combination with a TIG welding torch at a current of 50 A for welding aluminium alloy with a thickness of ≥ 3 mm. The penetration parameters and the process productivity achieved by the authors proved to be comparable with those obtained in welding using a 2 kW Nd:YAG-laser.

Benefits of laser-arc welding of thin-sheet aluminium and stainless steel are demonstrated in references [43, 45], which compare basic parameters of the three welding methods: TIG, combined and laser welding. At a laser power within the range of 0.1 to 1.5 kW (for the last two methods) and an arc length of up to 6 mm and 10 mm (for the first two methods), the typical welding speed was 0.5, 1.0 and 2.0 m/min, the penetration depth to weld width ratio was 0.5, 1.0 and 2.0 and heat input was 320, 160 and 50 J/mm, respectively. It is noted that the combined process is characterised by an improved arc burning stability and facilitated arc ignition. It is emphasised that the combined process can be employed for welding very thin sheets and joining elements of different thickness with no requirement for a high precision of fit-up of the joints for welding, as is required in the case of laser welding.

Positive results on combined welding of thin-sheet (0.5–1.0 mm) metals, including aluminium, but using plasma arc augmented laser welding (PALW) method* were obtained

---

* In classification of the combined processes suggested in Section 1.2 it can be referred to as laser + PA process.

by the authors of the earlier mentioned study [30]. Welding was performed using a CW CO_2-laser with a power of 400 W and a standard plasma torch operating at a current of up to 60 A and located as shown in Figure 1.10.

While substantiating the expediency of replacing the open arc by the plasma transferred arc in the combined process, the authors of [30] point out to a number of important drawbacks of the laser + GTA process, as a result of which it has not yet found wide commercial application, despite the potential capabilities noted by all investigators. Limitations of the method of laser + GTA welding are caused primarily by peculiarities of the open arc, which show up even in the presence of the laser beam and have a negative effect on characteristics of the combined process as a whole. These are the well-known difficulties associated with arc ignition and a spatial instability of the arc column.

As would be expected, the use of the constricted (plasma) arc in the combined process instead of the free-burning arc allowed the benefits of the former and this combination as a whole to be realised. The following important results were obtained: reliable arc ignition was provided (due to using a pilot arc) and service life of the tungsten electrode was increased. As far as the positive effects achieved by using this method, as compared with laser or plasma welding, are concerned, they are similar to those observed when using other types of the combined processes and have been considered above. In particular, this applies to the clearly defined synergistic effect which resulted in the formation of an increased volume of molten metal in laser + PA welding, as compared with the sum of the corresponding volumes of the melts in plasma and laser welding taken separately. Owing to this fact the authors of [30] succeeded in increasing the speed of welding thin-sheet metals by a factor of 2 to 3, as compared with purely laser welding. Thus, the full-penetration butt weld in stainless steel 0.6 mm thick was made at a laser power of 400 W, arc current of 50 A and a welding speed of 2500 mm/min, which was 2.5 times as high as that achieved when using the same power laser alone (400 W). The relationship between the plasma arc current and the speed of combined welding of stainless steel of the above thickness is illustrated in Figure 1.29. Full-penetration welds were successfully made also in a titanium alloy (Ti–6Al–4V) 0.75 mm thick at the same values of the process parameters.

Advantages of the PALW method are particularly evident in welding aluminium. The authors of [30] produced sound butt welds with full penetration in untreated aluminium 0.6 mm thick at a welding speed of 500 mm/min and low arc currents. The welding speed was restricted only by the power of the available laser. In this case, the anode spot of the arc

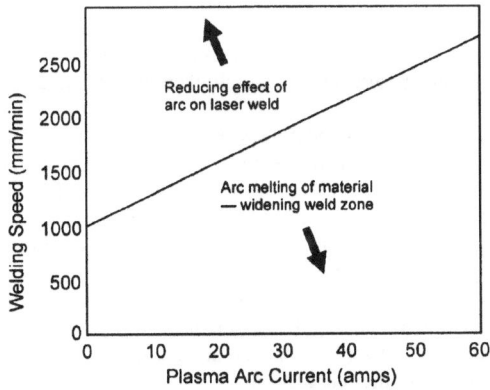

Figure 1.29 Relationship between plasma arc current and welding speed. Stainless steel (0.6 mm) butt welds [30].

that had formed within the limits of the metal heating zone made up under the effect of the laser beam moved together with the latter, which ensured continuity and full penetration of the weld. It should be noted that attempts made by the authors of this paper failed, in the first case because of wandering of the anode spot of the arc, and in the second case—because of insufficient power of the laser.

As noted in [30], other advantages of laser + PA welding, in comparison with laser welding, include the shape of the weld cross-section, which is more favourable in terms of strength of the welded joint and less stringent requirements to the accuracy of fit-up, and the value of the permissible deflection of the laser beam from the fusion line. Thus, unlike laser welding, where gaps between the mating elements should not exceed 5–10 % of the metal thickness, in combined welding gaps of up to 25–30 % may be tolerated.

It should be noted to complete the overview of the basic results of [30] that the investigations conducted allowed its authors to infer that the PALW method had wider capabilities in terms of commercial application than laser + GTA welding and that it would be necessary to have a deeper insight into the laser + PA processes, including those employed using high-power CO_2-lasers or solid-state Nd:YAG-lasers. Also, we should note the topicality of the task posed by the authors of [30], which consists in the development of specialised devices for commercial application of the PALW process, i.e., integrated laser heads or laser-arc plasma torches.

The problem of widening the fields of commercial application of the new laser-arc technologies, in particular their utilisation in the motor industry, was discussed at the International Conference IBEC-94 held in Detroit, USA [46]. In the presentation on the method of hybrid laser-arc welding sheet steel and aluminium realised by the combined effect of radiation of a Nd:YAG-laser and an arc with a non-consumable electrode, Beyer, a staff member of the Fraunhofer-Institute fuer Lasertechnik (Germany), noted a number of advantages of this process which were of particular importance for the motor industry, such as a decrease of 50 % in welding costs, a similar increase (50 %) in productivity, reduction in costs for edge preparation, possibility of controlling the weld width and, finally, maintaining the initial flexibility of a structure after welding.

While analysing in general the publications of the last decade, dedicated to combined laser-arc processes, it can be concluded that, instead of decreasing, the interest in these processes is gradually shifting from the sphere of experimental studies to that of commercial production. The focus of prospects for further development of laser-arc welding processes is included in the title of paper [32], which has been formulated by its authors as follows: "Synergic operation of welding arc and laser beam—for practical application or for scientific research only?" This paper, being one of the latest as to the time of its publication, is in fact an overview, which generalises and analyses achievements of scientists from different countries in the field of laser-arc processes.

The author of [32] gives a retrospective view of the history of laser-arc welding, describes advantages of the combined processes which have been extensively discussed above, and emphasises the synergistic effect of the combined use of laser radiation and the electric arc. After pointing out the drawbacks of laser TIG welding, the author comes to the analysis of papers on laser MIG and laser + PA welding and draws the conclusion that the latter methods are more promising for practical application in terms of the thickness ranges and the types of metals welded. As far as the laser TIG process is concerned, despite its drawbacks it will certainly find a wider application in the future, which is proved by positive experience gained from its utilisation (in particular, using a filler) for welding complex thin-walled structures and from other cases.

The author of [32] believes that the laser-arc processes are of interest in terms of both theoretical and experimental research, as there are many problems which are unsolved yet, such as those associated with the effect of wave length and laser radiation power on the

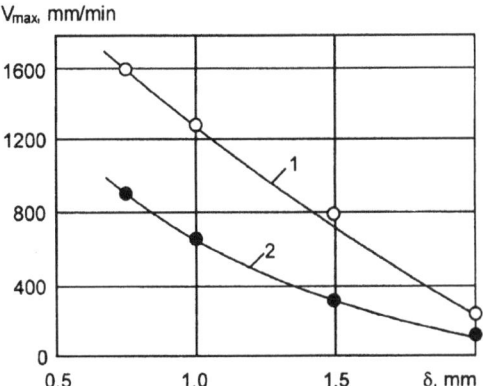

Figure 1.30 Maximum cutting speed against stainless steel plate thickness: 1, laser + PJ cutting (laser power 150 W, plasma nozzle diameter 1.5 mm, plasma jet input power 440 W); 2, laser cutting (laser power 150 W, spot size of focused laser beam 0.5 mm) [29].

degree of ionisation of the arc plasma at different arc currents. Therefore, he does not oppose the lines of development of combined welding as they are formulated in the title of the paper, nor does he choose between them. Instead, he highlights the prospects of both.

Despite the fact that in the overview considered we cannot find any references to studies conducted by scientists from the NIS countries, but this does not mean that no investigations of the laser-arc processes have been done there*. Let us begin the survey of these studies dedicated to investigation into operational capabilities of the combined processes for metal treatment from paper [48]. In particular, this paper highlights experience of applying laser + GTA welding for pipe production. The authors welded straight pipes 38 mm diameter of 08Kh18N10T steel 1.5 mm thick at speeds from 2.5 to 3.5 m/min. 2.5 kW CO_2-laser radiation was focused on a tail portion of the weld pool formed by a non-consumable electrode arc burning in argon at a direct current of (50–250 A). Better weld formation, especially in the upper part of the weld, was noted in laser-arc welding, as compared to laser welding, this allowing the requirements for groove preparation and butt fitting for welding to be reduced.

Paper [29] deals with the combined cutting of stainless steel plates up to 2 mm thick. Focused radiation of the solid-state laser and a non-transferred arc were used as components of the heat source (laser + PJ process). The authors used a CW Nd:YAG-laser with a power of about 200 W operating at the radiation wave length of 1.06 micron and a plasma torch of up to 1 kW power with a replaceable nozzle—an anode of the tungsten-electrode arc burning in a flow of argon. The laser cutter with air fed into its nozzle was placed over a plate at an angle of 60 degrees to its surface and the plasma torch—vertically under the plate.

The paper considered presents the results of experimental studies on the effect of the main parameters of laser + PJ cutting on its productivity. The experiments showed that laser-arc cutting allowed a substantial increase in the speed of treatment, as compared to laser cutting, without degradation of the cut quality. Thus, with 880 W power of the plasma arc (according to estimation done by the authors, it corresponds to the power introduced into a workpiece, i.e. 440 W) and with 150 W power of the laser radiation the speed increased approximately twice (Figure 1.30), but the cut width and the heat-affected zone sizes

---

* The first mention of the studies conducted in the former USSR in the field of laser-arc metal treatment methods refers to 1984 [47].

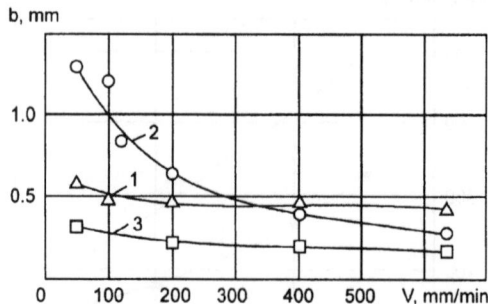

Figure 1.31 Width *b* of cut and heat affected zone vs. laser cutting speed of stainless steel plate (laser nozzle to workpiece distance 1.0 mm, laser power 150 W, spot size of focused laser beam 0.5 mm): 1, 3, width of cut on the face and back surfaces, respectively; 2, heating zone width [29].

remained close to the corresponding values typical for the laser process (Figures 1.31, 1.32). With an increase in the power of the plasma arc above the said value the speed of combined cutting did not grow and the cut parameters and quality became typical for the plasma process. The authors conclude that there exists an optimum ratio between the powers introduced by the plasma jet and the laser beam into a metal under treatment and estimate it to be not more than 3 : 1. However, while determining the heat input of the plasma jet into a workpiece, the authors take into account only the efficiency of metal heating by the plasma, forgetting, for example, the energy losses at the plasma torch anode. When allowance is made for these losses, the power introduced into a workpiece by the plasma jet will be considerably lower than 440 W. Therefore, in our opinion, the following conclusion will be more exact: for optimal laser-plasma cutting, the powers introduced into a metal by the plasma jet and the laser radiation should be approximately equal to each other. This statement is similar to that made in [8] for combined cutting using the transferred arc.

An attempt was made in [49–52] to evaluate the efficiency of the application of laser-arc heating for various kinds of heat treatment of materials, including surface strengthening, cladding, alloying, etc. For this a number of quantitative characteristics of the combined

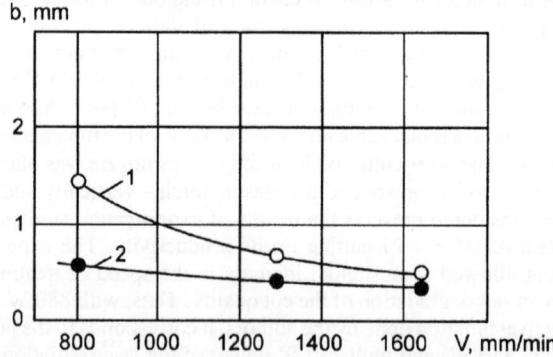

Figure 1.32 Width *b* of cut and heat influence zone vs. laser + PJ cutting speed of stainless steel plate (laser nozzle to workpiece distance 1.0 mm, laser power 150 W, plasma jet input power 440 W, plasma nozzle diameter 1.5 mm): 1, 2, width of heating zone and cut, respectively [29].

process are introduced, such as energy input coefficient $K_{la} = Q_a/Q_l$, where Q_a and Q_l are the powers of the arc and the laser beam, respectively, and the efficiency of the laser-arc action $\eta_{la} = (\eta_l Q_l + \eta_a Q_a) / (Q_l + Q_a)$, where η_l is the absorptivity of metal and η_a is the arc discharge efficiency. To evaluate the efficiency of the laser-arc metal treatment processes and to optimise them, the authors of the said papers suggest an investigation of the dependence of the η_{la} value on the energy input coefficient K_{la} for different but fixed values of η_l and η_a. The latter assumption, however, drastically decreases reliability of the results obtained under such an approach, since it does not allow for the basic peculiarity of the laser-arc effect on a workpiece, i.e. the mutual effect of the laser beam and the electric arc joined together in a combined process. As stated above, the arc discharge efficiency grows under the influence of the laser radiation [6, 8], while the additional arc heating of a treated metal increases the laser beam absorption efficiency [41]. The extent of this mutual effect depends not only on the ratio, but also on the absolute values of the powers of the laser radiation and the arc. Therefore, in quantitative estimation of the efficiency of the laser-arc metal treatment process, coefficients η_l and η_a should not be considered constant in varying the values of K_{la}, as well as Q_l and Q_a.

To complete the review of papers dealing with the energy characteristics and the technological peculiarities of laser-arc heat sources, the book by Duley [53] is noteworthy in describing the advances of laser technology over the past quarter of a century since the time the first laser appeared. By giving a brief analysis of laser-arc welding and cutting processes, the author emphasises two main aspects: firstly, the thermal effect of the combined laser-arc action exceeds the sum of the effects of each of the heat sources taken separately, and, secondly, the maximum increase in the process productivity is achieved with commensurate values of the arc and laser beam powers joined together in a combined heat source. Noting the high efficiency of laser-arc metal treatment methods, the author relates them to the prospects for further development of laser technologies.

1.4 Mutual Effects of the Laser Beam and the Electric Arc in Combined Processes

It can be seen from the review of the laser-arc methods for the joining and treatment of materials, that, along with the experimental studies conducted on the combined effects on workpiece, scientists, starting from Steen, made attempts to interpret the observed phenomena from the standpoint of the physics of the processes. A shift in emphasis towards elaboration of the physical concepts of the mechanism underlying the laser-arc processes took place in the second half of the eighties. In this respect, characteristic are the publications of the scientists from the former USSR, which appeared in periodicals, when the basic laser-arc effects on workpiece had already been described in the literature. Apparently, it is because of this fact that we can not find there detailed studies on the characteristics of metal melting and technological potentialities of combined treatment methods. These publications deal to a greater degree with the study of the mutual effects of laser radiation and electric-arc plasma to determine the optimum conditions for their combined action on materials.

For example, in [54–56] the main consideration is given to the effect of the laser beam on electric arc burning from a non-consumable electrode in an inert gas. Thus, paper [54] gives the results of experimental studies of the arc current-voltage characteristics in laser + GTA welding of 08Kh18N10T corrosion-resistant steel specimens in an argon atmosphere. The experiments were conducted by using the 1 kW CO_2-laser and the machine for tungsten electrode argon arc welding. In that case the arc torch was placed ahead of the beam at an angle of 45 degrees to the workpiece surface. The plots of arc voltage against the laser radiation power at various values of the welding current, as well as the arc current-voltage characteristics in the

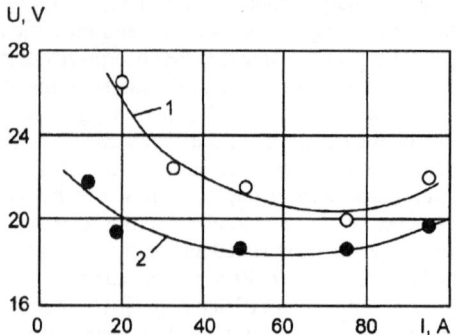

Figure 1.33 Current-voltage characteristics of arc: 1, GTA welding; 2, laser + GTA welding (laser power 900 W) [54].

arc and combined welding cases, were constructed by oscillograms of the arc gap current and voltage. The experiments showed, that over the entire range of the currents investigated the arc voltage in laser-arc welding was lower than in arc welding (Figure 1.33), this difference being greatest for laser beam power in excess of 500 W (Figure 1.34).

Later, the authors of [54] extended their studies of the effect of laser radiation on the electric parameters of the welding arc by using specimens of different materials [56]. Plots similar to those shown in Figure 1.34 were constructed for titanium, copper, aluminium and graphite. The results obtained prove that for all metals investigated the arc voltage in laser + GTA welding is lower than in arc welding (Table 1.2). An increase in the voltage under the influence of laser radiation was observed only for the arc burning with a graphite anode. However, the authors explain this fact by the increase in the effective arc length as a result of the narrow groove 2–3 mm deep formed at the graphite surface.

The decrease in the voltage of the arc with a metal anode which takes place under the influence of the laser beam with an effectively unchanging current (see Table 1.2) indicates an increase in conductivity of the arc gap. This, in fact, implies a growing concentration in it of charged particles. The main reason for an increase in density of charged particles in the arc plasma, from the viewpoint of the authors of [56], is the evaporation of the anode material by laser radiation. Also, the authors suppose that the observed decrease in the arc voltage is caused mostly by a decrease in the potential drop in the anode region. Admitting the important role of the evaporation mechanism in the decrease in the arc voltage taking place

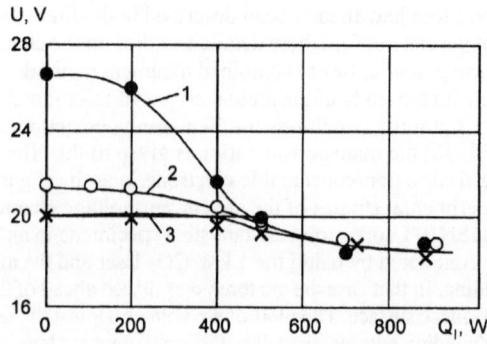

Figure 1.34 Arc voltage as a function of laser power: 1, arc current 20 A; 2, 50 A; 3, 75 A [54].

Table 1.2 Electrical parameters of the arc in arc and laser-arc welding of various materials [56].

Anode material	Arc welding		Laser-arc welding		
	Arc voltage (V)	Arc current (A)	Arc voltage (V)	Arc current (A)	Laser beam power (kW)
1. Corrosion resistant steel	26.6	20	20.2	20	0.5
2. Copper	29.9	75	24.8	76	0.9
3. Titanium	22.0	35	19.0	35	0.5
4. Aluminium	23.1	30	17.3	30	0.9
5. Graphite	17.0	47	27.0	45	0.9

in laser-arc welding, it is worth noting that the effect of the increase in plasma conductivity and, hence, the decrease in the arc gap voltage, can also be caused by additional argon arc plasma heating by the focused radiation of the CO_2-laser and can become evident even in the absence in it of anode material vapours. As to an assumption of a decrease in the anode potential drop, the known increase in the efficiency of the arc heating of metal in the presence of the laser beam, which is indicative of an increase in the volt equivalent of heat introduced by the arc into the anode, makes one doubt the correctness of this assumption. In our opinion, a decrease in the total arc voltage is caused, most probably, by a decrease in the intensity of the electric field in the discharge column due both to direct interaction of the laser radiation with the arc column plasma and the anode material vapours appearing in it with a lower ionisation potential than argon.

The authors of [54–56] dedicated a separate publication to the experimental study of the mutual effect of the laser beam and the electric arc as heat sources joined in the laser + GTA process [57]. For this purpose they studied laser, arc and combined laser-arc welding of stainless steel specimens in a helium atmosphere and, then, estimated the heat input into a workpiece for each of the above methods by taking cross-sectional areas of the welds. The mutual effect existing between the laser and arc heat sources in the combined process, which even Steen indicated in [6, 8], was experimentally proved by the fact that the heat input into a metal in laser-arc welding turned out to be higher than the sum of the heat inputs provided by arc and laser welding taken separately.

The total heat efficiency of the laser-arc process, η_{la}^t, determined as the ratio of the thermal power consumed for weld metal melting to the total power of the laser radiation and the arc discharge, was introduced to estimate quantitatively the degree of the mutual effect of the combined process heat source constituents. The dependence of η_{la}^t on the relative arrangement of the laser beam and the arc, and on the ratio of their powers were studied. As can be seen from the experimental data given in Table 1.3, the efficiency of the laser + GTA process is higher in the case when the arc torch is located ahead of the laser beam in the welding direction (all the results of work [57] considered below refer particularly to this case). As to the effect of the ratio of the laser radiation and electric arc powers on the

Table 1.3 Characteristics of the thermal effect on a workpiece in combined welding of stainless steel [57].

Relative positions of heat sources	Laser beam power (W)	Arc current (A)	Welding speed (mm/s)	Penetration area (mm²)	η_{la}^t
Arc ahead of laser beam	950	100	40	2.370	0.32
Arc behind laser beam	950	100	40	1.856	0.25

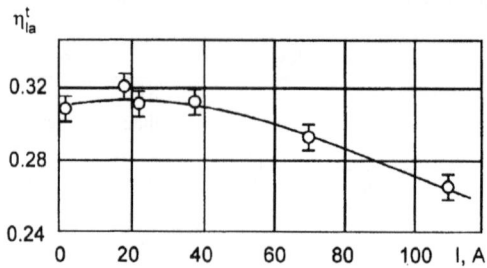

Figure 1.35 Total heat efficiency of laser + GTA welding against arc current (laser power 900 W, welding speed 21.3 mm/s) [57].

efficiency of the metal combined heating, the experiments showed that with fixed value of the arc current increase in laser radiation power from 100 to 1000 W caused an increase in the total process efficiency, as in the case of laser welding. In contrast, increase in the arc current from 15 to 120 A with constant power of laser radiation firstly (at low currents) almost does not affect the value of η_{la}^t, and then even results in its decrease (Figure 1.35). The authors explain this by the fact that the lower efficiency of the arc constituent of the heat source starts playing the main role in the decrease in the efficiency of the combined heating process with increase in the welding current above a certain limit. Most probably this is also the cause of the decrease in productivity of the combined process observed with increase in the arc power, as illustrated in Figure 1.36 (curve 1). This figure also gives the calculated dependence of the laser welding speed providing complete penetration of the specimen, on the laser radiation power (curve 2). Comparing these curves shows that while the arc power is lower than or approximately equal to the laser radiation power ($Q_1 = 900$ W), productivities of laser and laser-arc welding are approximately equal. With further increase in the arc power, the laser + GTA welding productivity decreases with respect to laser welding productivity, as in the case of laser-arc cutting (see Figure 1.4).

Papers [58, 59] stand out among the publications of other authors for the wide spectrum of studies conducted on the laser-arc effect on materials, covering both theoretical and applied aspects of the matter. An attempt was made in these papers to systemise and explain the phenomena occurring in combined welding, cutting and surface heat treatment, to reveal specific features of these processes against the corresponding laser technologies and deter-

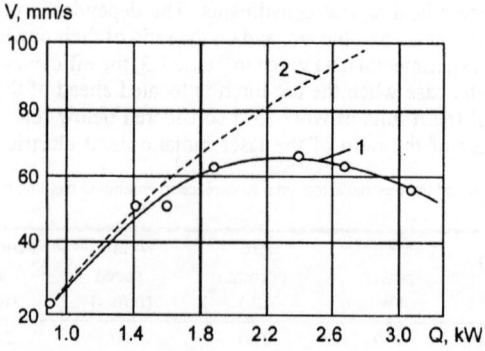

Figure 1.36 Maximum welding speed vs. total power of heat source: 1, laser + GTA welding (laser power 900 W); 2, laser welding [57].

mine the optimum conditions for the simultaneous use of laser radiation and electric arc in metal processing.

One of the objectives of the studies performed by the authors of [59] was to determine the role of polarity of the arc discharge electrodes in such an effect as the stabilisation of the arc root within the laser action zone in laser + GTA welding. In the experiments, the arc discharge was ignited between a metal being treated and a non-consumable electrode located on the same side of the workpiece as the laser beam at an angle of 45 degrees to the irradiation direction. The 700 W CW CO_2-laser served as a laser radiation source.

It was concluded on the basis of physical concepts of near-electrode processes taking place in the arc, as well as the experimental results given in [59], that polarity of the electrodes considerably affected stabilisation of the arc root. According to the authors' estimates, if a workpiece was a cathode, to make this effect manifest itself at rather high currents (tens of amperes) it was necessary to have radiation intensities requiring laser powers of the order of a few kilowatts. Parameters of the laser beam used for the experiments (power—600 W, diameter in focus—0.4 mm) did not provide the radiation intensity necessary for stabilisation of the cathode spot, and the location effect was not observed, i.e. the arc spot left the laser action zone and moved independently of the laser heating spot. Evidently, this case was not in fact one of combined effect of laser beam and arc on a metal and they can not be considered as a single thermal energy source.

If the workpiece is an anode, arc root stabilisation is of a threshold character. However, the threshold value of laser radiation intensity is much lower in this case and amounts to 10^5–10^6 W/cm^2. Since this value is close to the critical radiation intensity ensuring deep penetration mode in purely laser welding, the authors explain the location of the anode spot by a plasma-vapour jet with conductivity higher than that of the surrounding atmosphere forming over the zone of laser evaporation of metal. As we have already noted above, other researchers (see, for example, [8, 56]) are of the same opinion on the mechanism of arc anode spot stabilisation in the presence of laser radiation.

Searching for the optimum conditions of the laser-arc effect on a workpiece (anode), the authors of [59] studied the energy characteristics of the combined heat source and its components. Comparing the cross-sectional areas of welds made by laser, arc and laser-arc welding methods made it possible to determine for each of them the metal penetration rate and the process heat efficiency, η_{la}^m, determined as the ratio of the power consumed for melting of metal to the power absorbed by it. The values of melting rates for steel specimens and the heat efficiency for each heat source are given in Table 1.4. It follows from this Table that the melting rate in laser + GTA welding is almost twice as high as could be expected from simple addition of both components of the laser-arc source. This proves the strong mutual effect of the laser beam and the arc as heat sources joined together in a combined process.

Paper [59] deals also with the problem of the dynamic effect of the laser beam on the weld pool surface and its role in the change in the arc heat source characteristics during combined welding. Unlike the free-burning electric arc, which is practically a surface heat source, the focused laser radiation with power sufficient for intensive evaporation of metal

Table 1.4 Characteristics of the thermal effects of various heat sources on metal [59].

Thermal effect characteristics	Type of a heat source		
	CO_2-laser, 500 W power	Arc, 500 W power	Laser + arc, 500 + 500 W power
Metal melting rate (g/min)	0.80	6.00	14.0
η_m	0.05	0.18	0.27

is known to be a deep source. In a combined process the pressure of the yielding vapour jet forming as a result of laser evaporation of metal causes a sag of the weld pool surface, as in laser welding with deep penetration. Correspondingly, the arc heat source deepens into a workpiece. As a result, the laser-arc source acquires features typical of such a concentrated thermal energy source as a laser beam.

The high density of the power put into metal by the combined heat source allowed the authors of [59] to successfully apply laser-arc heating for cutting and surface hardening of various materials. For example, in the combined heat treatment of surfaces of structural steels and aluminium alloys (the process scheme with one-sided location of heat sources was used here, as in the case of welding) the authors managed to greatly increase the depth of the hardened layer as compared to laser treatment, without any increase in the laser power. In experiments on laser-arc cutting of thin-sheet heat-resistant alloys and composite materials, conducted by a method with the laser beam and the arc located on different sides of the workpiece, the authors came to results similar to those obtained in [8, 29]. In particular, the best cutting productivity and quality were achieved where the power put into the metal by each of the components of the combined heat source was approximately equal. Therefore the investigations conducted in [59] demonstrated that laser-arc cutting and surface modifying were promising applications as methods of metal processing, combining the high efficiency of the technological process and the possibility of using relatively low-power laser units.

The study [60] is dedicated to an important aspect of the problem of interaction taking place between the laser radiation and the electric arc in the combined metal treatment processes, in particular, to the effect of the arc discharge plasma on the laser beam propagating in it. The purpose of the authors was to investigate absorption of the laser radiation with different wave lengths by the plasma of the electric arc burning in an argon flow between a tungsten cathode and a copper water-cooled anode. The experiments used both a free-burning arc and an arc constricted by a gas flow (the degree of constriction was set by the flow rate of argon passing through a nozzle 2 mm diameter). The arc current varied from 40 to 300 A. The authors studied low-power laser beams with radiation wave lengths of 10.6 microns (3 W CO_2-laser) and 0.63 micron (1.5 mW He–Ne-laser), passing through the arc column in a direction normal to its axis.

Figure 1.37 shows the results of measuring transmissivity of the CO_2-laser radiation for the plasma of free-burning (curve 1) and constricted (curve 2) electric arcs as a function of current. It follows from the figure that the laser radiation transmissivity decreases with increase in the power and the degree of the arc constriction. The experimental data obtained allowed the authors to calculate, using the Lambert-Beer law, the 10.6 micron wave length laser radiation total absorptivity of the arc column plasma (the transverse size of the column necessary for the calculations and varying with current and gas flow rate was estimated visually). Dependencies of absorptivity on current for the free-burning and constricted arcs are given in Figure 1.38.

Study [60] also involved measuring the distributions of intensity of the probing CO_2-laser radiation in the beam cross-section* with and without the arc. It was found that the normal (Gaussian) distribution of the intensity along the radius in the initial laser beam had a gap in the central part oriented along the column axis, that formed after the beam had passed through the arc plasma. This is indicative of the optical heterogeneity of the arc column along the radius and is related to the higher concentration of the charged particles near the axis, which are responsible for absorption of the laser radiation by the arc plasma. As to the optical properties of the plasma varying along the height of the arc column, the authors note that the

---

* These experiments used a laser beam with a diameter comparable to the transverse size of the arc column.

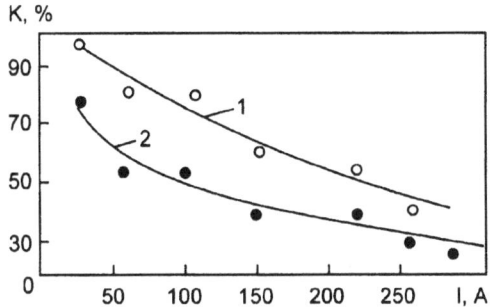

Figure 1.37 Transmissivity of arc column for laser radiation with wave length $\lambda = 10.6$ microns as a function of arc current: 1, free-burning arc; 2, constricted arc (arc length 8 mm, free-burning arc diameter 7–15 mm, argon rate 5.0 l/min) [60].

radiation transmissivity remains practically constant. Unfortunately, they give no details about the behaviour of the absorptivity.

The analogous experiments involving the He–Ne-laser radiation showed that the arcs, both free-burning and constricted, caused no attenuation (within the measurement error) of the laser beam passing through up to currents of 300 A. On the basis of this data, confirming the known dependence of the coefficient of absorption of the electromagnetic radiation by the plasma on its wave length, the authors conclude that the argon arc plasma is transparent for the 1.06 micron wave length Nd:YAG-laser radiation as well.

When characterising the study [60] as a whole, one should note that the results obtained by the authors can be helpful in estimating the effect of the arc plasma on the laser beam propagating in it, for different methods and conditions for realisation of the combined process. At the same time, one should bear in mind, that the low power of the probing radiation used by the authors in their experiments, excluded any effect of the laser beam on the arc plasma characteristics. In actual laser-arc processes, where the power introduced into the arc by the laser radiation is commensurate with the arc power, such an effect can result

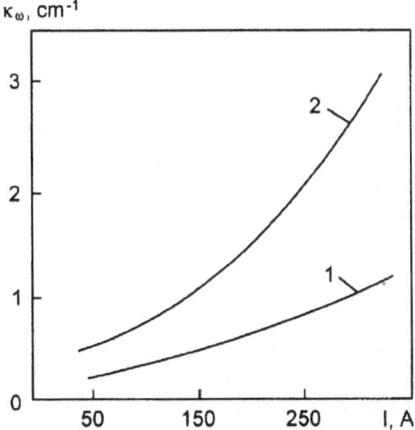

Figure 1.38 Absorptivity of laser radiation ($\lambda = 10.6$ microns) by electric arc as a function of arc current: 1, free-burning arc; 2, constricted arc [60].

in a substantial redistribution of the optical properties of the plasma and, hence, in a change in conditions of the laser beam propagating in it, as compared to the undisturbed arc plasma. Under certain conditions this may lead, as shown below, to the formation of new optical effects related to the self-action of the high-power laser beam in the combined discharge plasma.

Work [61] written by Chinese scientists is a rather detailed experimental study of the peculiarities of combined welding stainless steels using CO_2-laser radiation and a tungsten electrode argon arc (basic results of this study can also be found in a brief report [62]). Alongside with issues of a technological character, such as the determination of the metal melting parameters, estimation of the process efficiency and productivity, this work considers a wide spectrum of problems concerning the mutual effect of the laser beam and the electric arc in simultaneous use. Thus it was the first investigation (except for brief mentions in [41, 43, 45]) of the excitation of the arc discharge in the presence of a laser beam. In the experiments the arc torch was installed at an angle of 45 degrees to the radiation direction, so that the spacing between the electrode tip and the beam axis was 3–5 mm and that between the tip and the workpiece surface, 2–3 mm. 160 V voltage was supplied to the arc gap. After that the workpiece was irradiated with a 400 W CW laser. Filming at a speed of 50 frame/s showed that the arc was ignited at the moment when the laser plasma reached the non-consumable electrode, while expanding in the process of its development, i.e. to cover the entire discharge gap. With increase in the laser beam power, the reliability of the arc excitation was improved and the time interval from the moment the laser radiation was switched on until the moment of ignition of the arc discharge was reduced.

Paper [61] considers the experimental study of the mutual effect of the laser beam and the electric arc joined in a combined process with the location of the arc anode region in the laser plasma plume, resulting in a change in the shape of the arc discharge column. Referring to this phenomenon as the arc attraction effect, the authors note that its appearance depends on the arc current, the duration of stable burning of the arc before switching on the laser radiation, the distance of the electrode tip to beam axis, as well as on the laser power. The higher the current, the longer the stable burning, the larger the arc to beam axis distance and the higher the beam power — the smaller the effect of the plasma plume on the arc plasma column. In particular, with laser radiation power below 350 W, the arc current—above 30 A and with the electrode to beam distance of more than 3 mm, the attraction effect did not exist. The above regularities in the arc behaviour were observed with a workpiece both fixed as well as moving, the movement of the workpiece intensifying the attraction effect.

Stabilisation of the arc anode region within the laser heated zone of metal as a result of attraction of the arc to the laser plasma plume makes it possible to substantially increase the combined welding speed, as compared to that of arc welding. For example, in the said experiments arc current of 30 A and laser radiation power of 500 W remained stable up to a welding speed of 10 m/min, whereas in conventional GTA welding it lost its stability even at speeds of 2 m/min. Experimental dependencies between the ultimate speed of workpiece displacement and the arc current in laser + GTA and GTA welding methods obtained in [61] are similar to those presented in [41].

Based on the analysis of the results obtained in high-speed (3000 frame/s) filming of the arc behaviour in a combined process, the authors of [61] put the phenomena occurring during the interaction between the laser beam and the electric arc in the following sequence. After the switching on of the laser radiation, a cloud of metal vapour forms over the workpiece surface, and a high-temperature laser plasma is generated. The plasma plume, while expanding, attracts the arc column. This results in deflection of the latter and in the elongation of the arc. The near-anode zone of the arc column changes its colour (from red during independent burning to blue upon joining with the laser plasma plume), thus indicating an increase of the plasma temperature within the laser-arc interaction region.

In article [61] the following physical interpretation of the observed phenomena is given. Increase of the arc plasma temperature within the region of interaction with the high- temperature laser plasma leads to a growth of electrical conductivity and to a decrease in the electric field intensity in this region of the arc. Therefore, in compliance with the principle of minimum energy dissipation (Steenbeck's principle), the arc column tends to be formed within the limits of the laser plasma, while the anode spot tends to be located on the laser heating spot on the metal. In turn, stabilisation of the arc anode region in the laser plasma plume leads to an increase of the workpiece surface temperature (due to extra arc heating) and an increase in absorptivity of the metal within the zone of laser action. It should be noted here, that such an interpretation of the laser-arc interaction process wholly coincides with those suggested by other researchers in earlier studies (see, e.g. [6, 41]). The only new hypothesis brought forth by the authors of [61] as another probable cause of the increase in the efficiency of laser heating in the combined welding process, is decrease in density of the over-surface plasma as a result of its joining with the low-current (less than 90 A) arc plasma. A decrease in concentration of the charged particles in the plasma, and hence in its absorptivity, should result in an increase in the share of the laser beam energy reaching the metal surface.

The purpose of study [63] conducted later by the same authors was to confirm experimentally the hypothesis of decrease in density of the charged particles in the laser-arc plasma, as compared to the laser one. Here, the authors measured electron concentrations in the over-surface plasma in the cases of the laser, arc and the combined effect on a workpiece using the Stark method. The pulsed Nd-glass laser with radiation pulse energy of 5 J and pulse duration of less than 2 ms, as well as a TIG welding torch, were used for the experiments. An aluminium plate 0.8 mm thick served as the workpiece.

The results of measuring the electron density, n_e, (for laser plasma with radiation pulse energy of 5 J $n_e = 1.5 \times 10^{18}$ cm^{-3}; for arc plasma at a current of 30 A $n_e = 9.5 \times 10^{16}$ m^{-3}; for laser-arc plasma under the same conditions—$n_e = 5.2 \times 10^{17}$ cm^{-3}) formally confirm the assumption presented in [61] of a decrease in concentration of the charged particles in the over-surface plasma due to interaction of the focused CO_2-laser radiation with the arc discharge plasma. However, in our opinion, the data given cannot serve as experimental proof of the correctness of the said assumption, since they were obtained under totally different experimental conditions, i.e. when using a pulsed Nd-glass laser instead of a CW CO_2-laser. Besides, it was not indicated at what particular moment of time the electron concentration was measured with respect to the beginning of the laser radiation pulse. Meanwhile, it is common knowledge that parameters of the over-surface plasma greatly depend on wave length and, in the case of pulsed laser radiation, they vary with time. Therefore, the conclusion drawn by the authors of [63] of a decrease in the over-surface plasma density in the combined process is disputable with respect to the type of laser employed.

Paper [64] also refers to experimental studies on various aspects of the mutual effect of the laser beam and the arc in laser + GTA welding. The experiments used a process scheme with the heat sources arranged on the same side of the workpiece (a tungsten electrode torch was placed at an angle to the laser beam directed normally to the workpiece surface). The authors note a substantial increase in welding speed in the laser-arc process, an increase in penetration depth and its negligible dependence on the arc length and the actual value of the workpiece displacement speed. It is emphasised here, as well as in other studies, that the total thermal effect of the joined laser beam and arc exceeds the sum of the effects of each of the heat sources taken separately. Based on the results of measuring the metal melting parameters, observing the arc behaviour with the help of filming and the results of other experiments, the authors of [64] conclude that the laser radiation induced over-surface plasma plays a key role in the laser beam and arc interaction in the combined process. It promotes excitation, spatial stabilisation and contraction of the arc, i.e. it actually makes the laser beam controllable.

It follows from the review of the above studies that knowledge of the processes occurring during the interaction of focused laser radiation and electric arc plasma with each other and with the metal under treatment, has been increasing over time and has been supplemented with new experimental data. Accumulation of experimental information required a change in the lines and methods of investigation in gaining a deeper insight into the phenomena observed, their theoretical description and mathematical modelling.

Publications on theoretical investigations of the laser beam and electric arc plasma interaction as applied to laser-arc technological processes started appearing at the very end of the eighties. In particular, the results of a numerical study on the effect of CO_2-laser radiation on the characteristics of the arc discharge plasma inside a cylindrical channel were presented in [65, 66]. The problem was solved on the assumption that the plasma was homogeneous along the channel length and that it slightly affected the parallel laser beam propagating along the channel axis. It was shown that, under certain conditions, the system considered realised a special type of gas discharge (combined laser-arc discharge), wherein the radial distributions of temperature, plasma velocity and current density greatly differed from distributions of the corresponding characteristics of the conventional arc plasma inside a channel. In paper [67], an attempt was made to develop a mathematical model of the processes occurring in combined (laser + PA) welding. To make the model simpler, it was assumed that the arc plasma was axisymmetrical, and its basic characteristics, except for the natural current magnetic field and the radial velocity component, were averaged over the arc cross-section. Also it was assumed that the energy exchange between the laser beam and the arc plasma was due to inverse Bremsstrahlung absorption in the plasma. In this case the energy introduced by the laser beam was also averaged over the plasma arc cross-section.

Development of simplified models for the combined processes, based on various methods of averaging the parameters of the laser beam and the arc plasma allowed the main regularities of their mutual effects to be described, e.g. elevation of temperature and decrease in intensity of the electric field in the arc under the influence of the laser radiation, attenuation of the beam due to absorption of the laser radiation in the arc plasma, etc. More detailed study of the effects of the laser-arc interaction and the combined discharge peculiarities required the application of mathematical models with distributed parameters, i.e. models allowing for spatial distribution of characteristics of both the arc plasma and the focused laser beam interacting with it. Such studies were carried out at the E. O. Paton Electric Welding Institute and published in papers [68–71] dealing with the theoretical description and mathematical modelling of the laser-arc discharge initiated by the interaction between the focused CO_2-laser radiation and the plasma of the arc column. The studies conducted made it possible to predict new effects of the laser-arc interaction which may occur in the combined discharge, in addition to refining the concepts of the known regularities of the mutual effects of the laser beam and arc plasma. Taking into account the significance of the results obtained from the standpoint of the practical application of the combined discharge and the laser-arc plasma torches created on this basis to realise various technological processes, the authors think it reasonable to consider in detail the above papers in the next chapter.

CHAPTER 2

Laser Beam-Arc Plasma Interaction and Combined Discharge

In the numerous processes of laser-arc welding and treatment of materials considered in Chapter 1, the laser beam passes a certain distance in the arc column plasma. Interaction between the laser beam and the arc plasma occurring in this case can lead to a fundamental change in the energy balance of the arc discharge. The local character of heating the plasma by a focused laser beam induces substantial changes not only in the integrated, but also in the distributed, characteristics of the arc plasma. In addition, due to absorption and refraction of laser radiation in the plasma, there is redistribution of the intensity of the laser beam interacting with this plasma. If the power input into the arc provided by laser radiation is comparable with the electric power of the arc, a special type of discharge, i.e. combined laser-arc discharge, is realised [71]. The properties of this discharge differ greatly from the properties of a conventional arc discharge, and from properties of an optical discharge maintained by laser radiation.

Experimental and theoretical studies of the conventional electric arc are covered in many publications (see, for example, [72–75] and references mentioned in them). Different types of optical discharge maintained by laser radiation have been studied extensively [76–78]. At the same time, the peculiarities of the combined laser-arc discharge generated by interaction between a laser beam and arc plasma require special consideration. Therefore, the purposes of this chapter are the theoretical investigation of the process of interaction of a focused laser beam with arc plasma, the development of a mathematical model and detailed computer modelling of the combined discharge.

2.1 Physical Basis of Laser Beam and Arc Plasma Interaction and Mathematical Model of the Laser-Arc Discharge

2.1.1 General formulation

Consider a DC arc discharge affected by CW laser radiation under the following conditions. The arc is burning from a thermionic cathode in an axial flow of plasma-forming gas and, in its initial portion, is stabilised by the wall of a cylindrical channel with the focused laser beam propagating along its axis (Figure 2.1). The open region of the discharge (outside the channel) is subject to coaxial flow of a shielding gas at atmospheric pressure. The said conditions are realised, for example, in the device for laser-plasma processing of materials suggested in [26].

To construct a mathematical model for the discharge under consideration, assume that the annular thermionic cathode (or a system of pin cathodes arranged on the circumference) has an internal thin-walled cylindrical nozzle with radius R_1 (see Figure 2.1), aligned with the channel and forming with its wall an annular electrode nozzle ($R_1 \le r \le R_C$), through which a part of the plasma gas G_2 is fed and the total arc current I passes. Assume that the exit section of this nozzle is the beginning of the calculation domain of the discharge ($z = 0$), and suppose that it is located at a such distance from the cathode tip to allow us at the first stage to exclude the near-cathode processes from the description and consider the annular flow of the arc plasma at the cathode nozzle exit section to be axisymmetrical and one-dimensional in the direction of axis OZ, and thus to set correct input boundary conditions. Assume also that the rest of the plasma gas is fed through an internal cavity ($0 \leq r < R_1$) in the nozzle (see Figure 2.1) and enters the channel in a cold state, and that its flow rate G_1 can vary independently of the flow rate G_2 of the plasma gas fed through the electrode nozzle. Finally, assume that the arc anode is situated sufficiently far away from the cathode nozzle exit section and outside the calculation domain to allow us to neglect the anode processes in calculating the plasma characteristics of the discharge under consideration.

Physical phenomena occurring in such a system can be presented as follows. The thermal conductivity and viscosity processes result in increase of the temperature of the cold plasma gas axial flow, which becomes entrained into the arc plasma flow. As a consequence, at some distance Z_1 from the electrode nozzle exit section, the arc column collapses, i.e., fills in the entire channel section. Depending on radius R_C and length L_C of the channel, flow rates G_1 and G_2 of the plasma gas and arc current I, this will take place either in the closed ($Z_1 < L_C$) or open ($Z_1 > L_C$) parts of the discharge. The focused laser beam with the power Q_0 passes through the cathode nozzle opening and at $z = Z_1$ enters into the arc column (see Figure 2.1). Absorption of the laser radiation energy starts here, causing additional local (due to the small transverse sizes of the beam) heating of the arc plasma. If the power put into the plasma by the laser radiation is commensurate with the power generated in it by the electric current, key change in the characteristics of the arc plasma takes place in the laser-arc interaction region. The discharge under consideration is no longer an arc one, and a special type of discharge forms, i.e. a combined laser-arc discharge. Attenuation of the beam due to absorption and refraction in the plasma leads to a gradual decrease in its contribution to the energy balance of the combined discharge. As a result, starting from a certain section $z = Z_{II}$, where this contribution becomes sufficiently low, all the characteristics of the laser-arc discharge plasma again are restructured, approaching, in an asymptotical manner, those typical of a conventional arc in a gas flow.

Therefore, in theoretical analysis of the gas heating and motion processes taking place in the laser-arc discharge under consideration, the calculation domain of the discharge can be conventionally subdivided into three regions (see Figure 2.1):

I   the region of the arc column which does not interact with laser radiation ($0 \leq z < Z_1$);
II the region of the laser-arc interaction ($Z_1 \leq z \leq Z_{II}$), where combined discharge forms;
III the conversion region ($z > Z_{II}$), where all the characteristics of the combined discharge plasma relax to appropriate values typical of the fully developed arc flow.

2.1.2 Basic equations

For mathematical description of the processes occurring in interaction of the laser beam with the electric arc plasma column, assume the following:

1 the plasma system under consideration features cylindrical symmetry and the processes occurring are stationary;
2 the initial laser beam has axially symmetrical distribution of intensity (Gaussian beam);
3 the plasma is in a state of local thermodynamic equilibrium, and the natural plasma radiation is volumetric;
4 the main mechanisms of the plasma heating are Joule heat release and absorption of the laser radiation energy (work function of pressure forces and viscous dissipation can be ignored) and the energy transfer in the column is due to thermal conductivity and convection;

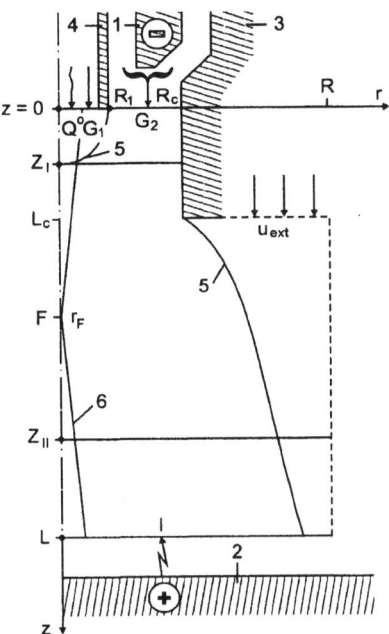

Figure 2.1 Schematic diagram of the laser-arc discharge: 1, cathode; 2, anode; 3, channel; 4, cathode nozzle; 5, arc boundaries; 6, laser beam boundary.

5 the plasma flow is viscous and subsonic; the mode of the flow is laminar; natural convection is ignored and external magnetic fields are absent.

Since in the discharge of the considered linear scheme the gas flows mostly in an axial direction, and the electric current density also has mostly an axial component, and while the radial temperature and velocity gradients are essentially higher than axial ones, a system of magnetic-gas dynamic (MGD) equations in the boundary layer approximation [75] can be used to calculate the plasma flow characteristics. With allowance made for the above assumptions, the conservation equations of energy and momentum and the continuity equation for the laser-arc discharge plasma can be written as follows [71]:

$$\rho C_p \left(v \frac{\partial T}{\partial r} + u \frac{\partial T}{\partial z} \right) = \frac{1}{r} \frac{\partial}{\partial r} \left(r \chi \frac{\partial T}{\partial r} \right) + \frac{j^2}{\sigma} + \kappa_\omega S - \psi \, ; \tag{2.1}$$

$$\rho \left(v \frac{\partial u}{\partial r} + u \frac{\partial u}{\partial z} \right) = \frac{1}{r} \frac{\partial}{\partial r} \left(r \eta \frac{\partial u}{\partial r} \right) - \frac{\partial}{\partial z} \left(p + \mu^0 \frac{H^2}{2} \right) \, ; \tag{2.2}$$

$$\frac{1}{r} \frac{\partial}{\partial r} (r \rho v) + \frac{\partial}{\partial z} (\rho u) = 0 \, , \tag{2.3}$$

where: $T(r, z)$ is the temperature; $u(r, z)$, $v(r, z)$ are, respectively, the axial and radial components of the plasma velocity; $p(r, z)$ is the pressure; ρ is the mass density; C_p is the specific heat at constant pressure; χ is the thermal conductivity coefficient; σ is the specific electrical conductivity; κ_ω is the laser radiation absorptivity; ψ is the volumetric radiative losses and η is the coefficient of the plasma dynamic viscosity.

Terms j^2/σ and $\kappa_\omega S$ in equation (2.1) describe, respectively, the energy release in the plasma due to electric current and absorption of the laser radiation [72, 76], where $j \equiv j_z (r, z)$ is the axial component of the current density in the discharge*, and $S (r, z)$ is the spatial distribution of the radiation intensity in the laser beam. Current density in the discharge can be found from Ohm's law:

$$j = \sigma E, \tag{2.4}$$

where $E \equiv E_z (z)$ is the axial component of the electric field. Within the framework of the boundary layer approximation used, the value of E is constant over the discharge section and can be determined from the condition of the total current conservation:

$$I = 2\pi E \int_0^{R_\sigma} \sigma r \, dr, \tag{2.5}$$

where $R_\sigma (z)$ is the radius of the current-conducting region. Taking into account that outside this region the plasma conductivity is practically equal to zero, the radius of the calculation domain can be used as the upper limit of integration in formula (2.5), i.e.: $R_\sigma (z) = R_C$ at $0 \leq z \leq L_C$ and $R_\sigma (z) = R$ at $z > L_C$ (see Figure 2.1).

The distribution of pressure $p (r, z)$ in the channel $(0 \leq z \leq L_C)$ is calculated allowing for magnetic pressure, and gives:

$$p = p_{ext} - \int_z^{L_C} \frac{dp_C}{dz} dz + \mu^0 E \int_r^{R_C} \sigma H \, dr, \tag{2.6}$$

where: p_{ext} is the gas pressure in the environment; μ^0 is the universal magnetic constant; $H \equiv H_\varphi (r, z)$ is the azimuthal component of the magnetic field due to the arc current:

$$H = \frac{1}{r} \int_0^r jr \, dr. \tag{2.7}$$

The gradient of gas static pressure $dp_C / dz (z)$ in the boundary layer approximation is constant over the channel section and can be found from the condition of conservation of the total plasma gas flow rate:

$$G = 2\pi \int_0^{R_C} \rho u r \, dr, \tag{2.8}$$

where $G = G_1 + G_2$ is the mass gas flow rate through the channel (see Figure 2.1). In the open region of the discharge $(z > L_C)$ the pressure is determined from the following relationship:

---

* Significant axial gradients of the electromagnetic characteristics of plasma forming in the initial sections of the channel can lead to values of the radial electric current component $j_r = - \partial H/\partial z$, commensurate with j_z, and this component should be taken into account in the Joule source j^2/σ of equation (2.1) along with the conservation equation of momentum (2.2).

$$p = p_{ext} + \mu^0 E \int\limits_{r}^{R} \sigma H \, dr .$$

(2.9)

The system of equations from (2.1) to (2.9) should be supplemented with the following expressions:

$$\rho = \rho\,(T,p); \quad C_p = C_p\,(T,p);$$

$$\chi = \chi\,(T,p); \quad \eta = \eta\,(T,p); \quad \sigma = \sigma\,(T,p);$$

$$\kappa_\omega = \kappa_\omega\,(T,p); \quad \psi = \psi\,(T,p),$$

(2.10)

determining the dependences of thermodynamic parameters, transport coefficients and optical properties of plasma on temperature and pressure.

To complete the basic set of equations, it is necessary to determine the distribution of the radiation intensity $S\,(r, z)$ in the laser beam by taking into consideration its refraction and absorption in the discharge plasma. Suppose that the relative variation in the plasma parameters at distances of the order of a wave length of the laser radiation is small, and that there is no laser radiation reflected from the anode surface. To describe the propagation of the beam in the heterogeneous plasma, we will use the quasi-optical approximation [79, 80]. Assume that the laser beam has the form of a quasi-plane electromagnetic wave, wherein the energy flow is directed preferably along axis OZ (Figure 2.1), and consider that the electric field vector of the wave has only a transverse component, then:

$$E_\omega = e \, Re \left\{ A_\omega \exp \left[i \, (kz - \omega t) \right] \right\},$$

(2.11)

where: e is the unit vector of the radiation polarisation; $A_\omega\,(r, z)$ is the slowly varying complex amplitude of the field; $k \equiv k_z = 2\pi/\lambda$ is the wave vector; ω is the angular frequency and λ is the wave length of the laser radiation in the absence of plasma. In the case under consideration the parabolic equation for the amplitude of electric field of the laser beam propagating in plasma will be written as follows:

$$- 2ik \frac{\partial A_\omega}{\partial z} = \frac{1}{r} \frac{\partial}{\partial r} \left(r \frac{\partial A_\omega}{\partial r} \right) + k^2 \, (\varepsilon_\omega - 1) \, A_\omega ,$$

(2.12)

and its solution should be found allowing for the following dependence:

$$\varepsilon_\omega = \varepsilon_\omega\,(T,p),$$

(2.13)

where $\varepsilon_\omega = \varepsilon_\omega' + i\varepsilon_\omega''$ is the complex dielectric permittivity of the plasma at the frequency of laser radiation. The value of $S\,(r, z)$ to be determined is actually the time average of the axial component of the electromagnetic energy density flow and is related to the complex amplitude of the electric field of the wave by relationship [81]:

$$S = \frac{1}{2} \sqrt{(\varepsilon^0/\mu^0)} \, |A_\omega|^2,$$

(2.14)

where ε^0 is the vacuum dielectric permittivity. We thus have a closed system of equations describing the effect of the laser beam on the electric arc column plasma taking into account the effect of plasma on the radiation beam propagating in it.

2.1.3 Boundary conditions

To solve the system of parabolic equations (2.1) to (2.3), (2.12), it is necessary to set the appropriate boundary and initial (input) conditions at the boundaries of the calculation domain $\{0 \leq r \leq R_C$ at $0 \leq z < L_C$; $0 \leq r \leq R$ at $L_C \leq z < L\}$ (see Figure 2.1). The boundary conditions at the discharge axis are selected on the basis of an assumption that the system features cylindrical symmetry, i.e. at $r = 0$; $0 \leq z < L$:

$$\frac{\partial T}{\partial r} = 0; \quad \frac{\partial u}{\partial r} = 0; \quad v = 0; \quad \frac{\partial A_\omega}{\partial r} = 0 .$$

$$(2.15)$$

It is assumed that at the external boundary of the calculation domain within the channel the temperature of the plasma gas is equal to the temperature of the cooled wall, T_C, and the condition of "sticking" is met, i.e. at $r = R_C$; $0 \leq z < L_C$:

$$T = T_C; \quad u = 0.$$

$$(2.16)$$

Supposing also that in any section at z ($0 \leq z < L$) the transverse size of the beam is less than the channel radius, the value of R_C can be selected as a radius of the calculation domain for equation (2.12) and it can be considered that at $r = R_C$; $0 \leq z < L$

$$A_\omega = 0.$$

$$(2.17)$$

The conditions of smooth contingency with an environment are assumed to exist at the external boundary of the calculation domain for the discharge in open region; i.e. at $r = R$; $L_C \leq z < L$:

$$T = T_{ext}; \quad u = u_{ext} .$$

$$(2.18)$$

Here: T_{ext}, u_{ext} are the values of temperature and velocity of the external gas flow, and the R value is selected such that at any z ($L_C < z < L$)

$$\left. \frac{\partial T}{\partial r} \right|_{r = R} = 0\,(\alpha); \quad \left. \frac{\partial u}{\partial r} \right|_{r = R} = 0\,(\alpha),$$

$$(2.19)$$

where $0\,(\alpha)$ implies equality to zero accurate to a certain preset small value of α. Finally, the conditions of (2.18) will also be used at $z = L_C$; $R_C \leq r < R$.

Suppose that at the inlet section of the channel, i.e. at $0 \leq r \leq R_C$; $z = 0$

$$T = T(r, 0); \quad u = u(r, 0); \quad v = 0; \quad A_\omega = A_\omega(r, 0).$$

$$(2.20)$$

Assume the following for determining the explicit form of distributions $T(r, 0)$ and u $(r, 0)$. The homogeneous isothermal flow of the cold plasma gas with temperature T_1 enters the channel through an internal cavity in the cathode nozzle. The nozzle wall is assumed to be infinitely thin and porous near the exit section. Then, at $0 \leq r \leq R_1$:

$$T(r, 0) = T_1; \quad u(r, 0) = u_1 \equiv \frac{G_1}{\pi R_1^2 \rho_1}.$$

$$(2.21)$$

where ρ_1 is the gas density at temperature T_1. Distributions of the plasma temperature and velocity at the electrode nozzle exit section, i.e. at $R_1 \leq r \leq R$, will be found by solving the one-dimensional equations:

$$\frac{1}{r}\frac{d}{dr}\left(r\chi\frac{dT}{dr}\right) + \sigma E^2 - \psi = 0;$$

(2.22)

$$\frac{1}{r}\frac{d}{dr}\left(r\eta\frac{du}{dr}\right) - \frac{dp}{dz} = 0,$$

(2.23)

where the values of $E(0)$; $dp/dz(0)$ are determined from the integral relationships:

$$I = 2\pi E \int_{R_1}^{R_C} \sigma r\, dr;\ G_2 = 2\pi \int_{R_1}^{R_C} \rho u r\, dr.$$

(2.24)

The boundary conditions for equations (2.22), (2.23) are selected in compliance with conditions (2.16), (2.21), at $r = R_1$, giving:

$$T = T_1;\ u = u_1,$$

(2.25)

and, at $r = R_C$:

$$T = T_C;\ u = 0.$$

(2.26)

To assign the initial distribution of $A_\omega(r, 0)$ assume that the focused Gaussian beam (TEM$_{00}$ mode) of laser radiation, which has minimum radius r_F in the plane $z = F$ (see Figure 2.1), is introduced into the discharge through the nozzle opening. The spatial distribution of the complex amplitude of the field of such a beam, satisfying equation (2.12) at $\varepsilon_\omega = 1$, will be determined by expression [80]:

$$A_\omega = A_{\omega F}\frac{r_F}{r_z}\exp\left[-\frac{r^2}{r_z^2} + i\left(k\frac{r^2}{2R_z} - \varphi_z\right)\right].$$

(2.27)

Here:

$$r_z^2 = r_F^2\left[1 + \frac{(z - F)^2}{z_F^2}\right];$$

(2.28)

$$R_z = (z - F)\left[1 + \frac{z_F^2}{(z - F)^2}\right];$$

(2.29)

$$\varphi_z = \operatorname{arctg}\left(\frac{z - F}{z_F}\right),$$

(2.30)

where

$$z_F = \frac{kr_F^2}{2}.$$

(2.31)

The constant $A_{\omega F}$ in (2.27) is found from the integral relationship for the total radiation power:

$$Q = 2\pi \int_0^\infty Sr\, dr,$$

(2.32)

with allowance for (2.14) giving:

$$A_{\omega F} = \sqrt{(4Q^0 / \pi r_F^2) \, (\mu^0 / \varepsilon^0)^{1/2}} \,, \tag{2.33}$$

where $Q^0 = Q(0)$ is the power of the initial laser beam. Allowing for the small transverse size of the beam, as compared to the channel radius, we will use the value R_C as the upper limit of integration in (2.32). Requiring that the transverse size of the beam at the channel inlet is smaller than the nozzle radius, the distribution of the complex amplitude of the field at $z = 0$; $0 \le r \le R_1$ can be determined from formulae (2.27) to (2.31), (2.33). The relationship obtained can be supplemented with condition $A_\omega(r, 0) = 0$ at $R_1 \le r \le R_C$.

2.1.4 Solution technique

The system of equations (2.1) to (2.3), (2.12), as well as the boundary conditions (2.15) to (2.18) and relationships (2.4) to (2.9) and (2.14) allow complete determination of the temperature and velocity of the plasma, the electric field intensity and other characteristics of the arc affected by laser radiation and the spatial distribution of the intensity of the laser beam interacting with it. The possibilities of finding analytical solutions for such systems of non-linear differential equations are rather limited. Therefore the problem was solved numerically by the finite-difference method using the basic differential scheme for integrating the system of equations of boundary layer type [82]. A similar approach was used, for example, for numerical simulation of gas heating and acceleration processes in conventional electric arcs [75].

The second-order differential equations (2.1), (2.2), (2.12) were approximated by the implicit two-layer six-point differential scheme, while the first-order equation (2.3)—by the explicit four-point one. The algebraic system of equations obtained was solved by the run method using iterations. The method of lower relaxation [83] was used to improve contingency of the iteration procedure. Stability of the differential scheme employed was studied by varying the grid mesh, which was selected such that the use of a finer mesh did not affect the numerical solution obtained within the preset accuracy.

The conditions for the integral conservation of energy and momentum of the plasma flow were checked to the preset level of accuracy for each layer z. Within the channel ($0 \le z < L_C$):

$$W = W(0) + 2\pi \int_0^z \int_0^{R_C} \left(\frac{j^2}{\sigma} + \kappa_\omega S - \psi \right) r \, dr \, dz - 2\pi R_C \int_0^z q_C \, dz \,, \tag{2.34}$$

where

$$W(z) = 2\pi \int_0^{R_C} \rho u h \, dr$$

is the flow of enthalpy $h = \int_{T_C}^{T} C_p \, dT$ through the discharge cross-section and

$$q_C = - \chi \frac{\partial T}{\partial r}, \text{ at } r = R_C$$

is the conductive heat flow from plasma to the channel wall;

$$K = K(0) - 2\pi R_C \int_0^z \tau_C \, dz ,$$

(2.35)

where

$$K(z) = 2\pi \int_0^{R_C} \left(p + \rho u^2 + \mu^0 \frac{H^2}{2} \right) r \, dr$$

is the total pulse flow and

$$\tau_C(z) = - \eta \frac{\partial u}{\partial r}, \text{ at } r = R_C$$

is the viscous friction at the channel wall.

In the open region of the discharge ($L_C \leq z \leq L$), making allowance for the conditions of smooth contingency with the environment (2.18), (2.19), we have for W and K:

$$W = W(L_C) + 2\pi \int_{L_C}^z \int_0^R \left(\frac{j^2}{\sigma} + \kappa_\omega S - \psi \right) r \, dr \, dz ,$$

(2.36)

where

$$W(z) = 2\pi \int_0^R \rho u h \, dr$$

and

$$K = K(L_C),$$

(2.37)

where

$$K(z) = 2\pi \int_0^R \left(p + \rho u^2 + \mu^0 \frac{H^2}{2} \right) r \, dr .$$

The condition of the integral energy balance between the laser beam and the plasma was controlled as well:

$$Q = Q(0) - 2\pi \int_0^z \int_0^{R_C} \kappa_\omega S r \, dr \, dz.$$

(2.38)

Here $Q(z)$ is the beam power in section z determined from relationship (2.32).

2.2 Thermal, Transport and Optical Properties of the Combined Discharge Plasma

Prior to numerical modelling of the laser-arc discharge, it is necessary to determine the dependences of thermophysical parameters, transport coefficients and optical properties of the discharge plasma on the temperature and pressure. Since in the majority of the known methods of laser + GTA and laser + PA welding and heat treatment such inert gases as Ar or He (or their mixtures) are used as a plasma-forming medium, we will study the above properties for an Ar – He plasma for an arbitrary ratio of the initial components.

When determining the temperature dependence of the equilibrium composition of the Ar – He plasma in the interval of $300\ \text{K} \le T \le 25000\ \text{K}$, it is taken to be a non-ideal six-component ($\alpha = 1, 6$) mixture containing e ($\alpha = 1$); Ar ($\alpha = 2$); Ar^+ ($\alpha = 3$); Ar^{++} ($\alpha = 4$); He ($\alpha = 5$); He^+ ($\alpha = 6$) in a state of local thermodynamic equilibrium. Then the system of equations for the calculation of the particle concentrations in such plasma will include the ionisation balance equations, the equation of state, the condition for conservation of the ratio between the plasma-forming mixture components and the quasi-neutrality condition [84]:

$$\frac{n_e n_{\alpha+1}}{n_\alpha} = \frac{2\theta_{\alpha+1}}{\theta_\alpha} \left(\frac{m_e k_B T}{2\pi\hbar^2}\right)^{3/2} \exp\left(-\frac{U_\alpha - \Delta U_\alpha}{k_B T}\right), \quad (\alpha = 2, 3, 5)$$

(2.39)

$$n \equiv \sum_{\alpha=1}^{6} n_\alpha = \frac{p}{k_B T},$$

(2.40)

$$\sum_{\alpha=5}^{6} n_\alpha \Big/ \sum_{\alpha=2}^{6} n_\alpha = \delta;$$

(2.41)

$$\sum_{\alpha=1}^{6} Z_\alpha n_\alpha = 0,$$

(2.42)

where: $n_1 = n_e$ is the concentration of electrons; n_α; θ_α ($\alpha = \overline{2, 6}$) are the concentrations and the statistical sums of atoms and ions; m_e is the electron mass; k_B is the Boltzmann constant; $\hbar$ is the Plank constant divided by 2π; $U_2 = U_{\text{Ar}}$; $U_3 = U_{\text{Ar}^+}$ are the ionisation potentials of atoms and singly charged ions of argon, respectively; $U_5 = U_{\text{He}}$ is the helium atom ionisation potential; $\Delta U_\alpha = (Z_\alpha + 1)\, e^2/4\pi\varepsilon^0 r_D$ ($\alpha = 2, 3, 5$) are the reductions in the ionisation potentials caused by the interaction between the charged particles in the plasma; Z_α is the charge number ($Z_1 = -1$; $Z_2 = Z_5 = 0$; $Z_3 = Z_6 = 1$; $Z_4 = 2$); e is the elementary charge;

$$r_D = \sqrt{\varepsilon^0 k_B T \Big/ \sum_{\alpha=1}^{6} (Z_\alpha e)^2 n_\alpha}$$

is the Debye radius, and δ is the parameter determining the volume content of He in the initial mixture.

Knowing the temperature and pressure dependence of the Ar – He plasma composition, with the help of formulae:

$$\rho = \sum_{\alpha=1}^{6} m_\alpha n_\alpha;$$

(2.43)

$$C_p = \left(\frac{\partial h}{\partial T}\right)_p$$

(2.44)

it will not be difficult to calculate the density and heat capacity of this plasma at constant pressure. Here: $m_1 = m_e$; $m_2 \approx m_3 \approx m_4 = M_{\text{Ar}}$ is the argon atom mass; $m_5 \approx m_6 = M_{\text{He}}$ is the helium atom mass; $h\,(T,p)$ is the specific enthalpy of the plasma determined with allowance for the ionisation energy [85]:

$$h = \frac{1}{\rho}\left[\frac{5}{2}k_B nT + k_B T^2 \sum_{\alpha=2}^{6} n_\alpha \frac{\partial \ln(\theta_\alpha)}{\partial T} + \varepsilon_i\right],$$ (2.45)

where $\varepsilon_i = n_3(U_2 - \Delta U_2) + n_4(U_2 - \Delta U_2 + U_3 - \Delta U_3) + n_6(U_5 - \Delta U_5)$.

The system of algebraic equations (2.39) to (2.42) was solved numerically by the Newton method. Numerical differentiation was used to calculate $h\,(T,p)$ and $C_p(T,p)$. The data on the energy levels of atoms and ions from [86] were employed to calculate the statistical sums θ_α. As an example, Figures 2.2 and 2.3 show the results of calculation of $\rho(T)$ and $C_p\,(T)$ for the atmospheric pressure Ar – He plasma with different helium contents of the initial mixture. The results obtained are in good agreement with the experimental and calculated data by other authors both for pure gases [87– 89] and for their mixtures [90].

The results of [91] obtained on the basis of solving the system of kinetic equations for the multicomponent plasma by the Grad method [92] can be used to calculate the transport coefficients $\eta\,(T,p)$, $\chi\,(T,p)$, $\sigma\,(T,p)$ for the plasma of a known composition. In particular, expressions for viscosity coefficient, transport thermal conductivity* and electrical conductivity of the equilibrium Ar – He plasma in the 13-momentum approximation by the Grad method have the following form [91]:

$$\eta = \sum_{\alpha=1}^{6} \frac{n_\alpha}{n} \sum_{\beta=1}^{6} \frac{n_\beta}{n} \frac{|N|_{\beta\alpha}}{|N|};$$ (2.46)

$$\chi^{tr} = \sum_{\alpha=1}^{6} \frac{n_\alpha}{n} \sum_{\beta=1}^{6} \frac{n_\beta}{n} \frac{|\Lambda|_{\beta\alpha}}{|\Lambda|};$$ (2.47)

$$\sigma = \frac{[\sigma_e]_1}{1-\Delta},$$ (2.48)

where: $|N|$ and $|\Lambda|$ are the determinants made up of coefficients $N_{\alpha\beta}$ and $\Lambda_{\alpha\beta}$; $|N|_{\beta\alpha}$ and $|\Lambda|_{\beta\alpha}$ are the algebraic adjuncts of elements $\beta\alpha$ of the corresponding determinants. Coefficients $N_{\alpha\beta}$, $\Lambda_{\alpha\beta}$ ($\alpha = 1, 6$) and the Δ value are determined by relationships [91]:

$$N_{\alpha\alpha} = \frac{n_\alpha^2}{n^2[\eta_{\alpha\alpha}]_1} + \sum_{\beta\neq\alpha} \frac{2n_\alpha n_\beta}{n^3(m_\alpha + m_\beta)[D_{\alpha\beta}]_1}\left(1 + \frac{3}{5}\frac{m_\beta}{m_\alpha}A_{\alpha\beta}\right);$$

$$N_{\alpha\beta} = -\frac{2n_\alpha n_\beta}{n^3(m_\alpha + m_\beta)[D_{\alpha\beta}]_1}\left(1 - \frac{3}{5}A_{\alpha\beta}\right), \quad \beta \neq \alpha;$$ (2.49)

$$\Lambda_{\alpha\alpha} = \frac{n_\alpha^2}{n^2[\chi_{\alpha\alpha}]_1} + \frac{4}{25}\frac{T}{P}\sum_{\beta\neq\alpha}\frac{n_\alpha n_\beta}{n^2(m_\alpha + m_\beta)^2[D_{\alpha\beta}]_1}\times$$

$$\times\left(\frac{15}{2}m_\alpha^2 + \frac{25}{4}m_\beta^2 - 3m_\beta^2 B_{\alpha\beta} + 4m_\alpha m_\beta A_{\alpha\beta}\right);$$

---

* To calculate the total thermal conductivity of plasma, in addition to χ^{tr}, it is necessary to also take into account the ionisation thermal conductivity χ^{ion} [93] caused by the ionisation energy transfer.

$$\Lambda_{\alpha\beta} = -\frac{4}{25}\frac{T}{p}\frac{m_\alpha m_\beta}{(m_\alpha + m_\beta)^2}\frac{n_\alpha n_\beta}{n^2[D_{\alpha\beta}]_1}\left(\frac{55}{4} - 3B_{\alpha\beta} - 4A_{\alpha\beta}\right),\quad \beta \neq \alpha\,;$$

(2.50)

$$\Delta = \frac{\left[\displaystyle\sum_{\beta=2}^{6}\frac{n_\beta}{[D_{e\beta}]_1}\left(\frac{6}{5}C_{e\beta} - 1\right)\right]^2}{\left[\dfrac{8}{25}\dfrac{n_e}{[D_{ee}]_1} + \displaystyle\sum_{\beta=2}^{6}\dfrac{n_\beta}{[D_{e\beta}]_1}\left(1 - \dfrac{12}{25}B_{e\beta}\right)\right]\displaystyle\sum_{\beta=2}^{6}\dfrac{n_\beta}{[D_{e\beta}]_1}}.$$

(2.51)

Here:

$$[\sigma_e]_1 = \frac{3n_e e^2}{16\,m_e \displaystyle\sum_{\beta=2}^{6} n_\beta\,\Omega_{e\beta}^{(1,1)}}$$

is the electron electrical conductivity;

$$[D_{\alpha\beta}]_1 = \frac{3k_B T}{16n\mu_{\alpha\beta}\,\Omega_{\alpha\beta}^{(1,1)}}$$

is the mutual diffusion coefficient of particles of the α and β kinds;

$$[\eta_{\alpha\alpha}]_1 = \frac{5k_B T}{8\,\Omega_{\alpha\beta}^{(2,2)}};\quad [\chi_{\alpha\alpha}]_1 = \frac{15}{4}\frac{k_B}{m_\alpha}[\eta_{\alpha\alpha}]_1$$

are the viscosity coefficient and the thermal conductivity of a simple gas consisting of the α kind particles, calculated in the first approximation by the Chapman-Enskog method [94]; $\mu_{\alpha\beta} = m_\alpha m_\beta/(m_\alpha + m_\beta)$ is the reduced mass. The values of $A_{\alpha\beta}$, $B_{\alpha\beta}$ and $C_{\alpha\beta}$ are determined by expressions:

$$A_{\alpha\beta} = \frac{\Omega_{\alpha\beta}^{(2,2)}}{2\Omega_{\alpha\beta}^{(1,1)}};\quad B_{\alpha\beta} = \frac{5\Omega_{\alpha\beta}^{(1,2)} - \Omega_{\alpha\beta}^{(1,3)}}{3\Omega_{\alpha\beta}^{(1,1)}};\quad C_{\alpha\beta} = \frac{\Omega_{\alpha\beta}^{(1,2)}}{3\Omega_{\alpha\beta}^{(1,1)}},$$

(2.52)

where $\Omega_{\alpha\beta}^{(l,s)}$ are the generalised integrals of Chapman-Cowling [94]:

$$\Omega_{\alpha\beta}^{(l,s)} = \left(\frac{2\pi k_B T}{\mu_{\alpha\beta}}\right)^{1/2}\int_0^\infty\int_0^\pi w_{\alpha\beta}^{2s+3}\,e^{-w_{\alpha\beta}^2}(1 - \cos^l\vartheta)\,\sigma_{\alpha\beta}(w_{\alpha\beta},\vartheta)\sin\vartheta\,d\vartheta\,dw_{\alpha\beta}$$

(2.53)

$\sigma_{\alpha\beta}$ is the differential cross-sections of shape-elastic scattering;

$$w_{\alpha\beta} = \sqrt{\mu_{\alpha\beta}(\mathbf{v}_\alpha - \mathbf{v}_\beta)^2/2k_B T}$$

is the dimensionless relative velocity of colliding particles and ϑ is the scattering angle.

The viscosity coefficient, thermal and electrical conductivities of the equilibrium Ar – He plasma for the temperature range from 300 K to 25000 K were calculated on the basis of formulae (2.46) to (2.52). The data on differential cross-sections of the scattered particles from [95–98] were used to calculate collision integrals. The T dependences of η, χ and σ at atmospheric pressure and with different helium contents of the mixture are shown in Figures 2.4 to 2.6. The results of calculations of the transport properties for the Ar and He plasma are in good agreement with the available experimental and calculation data obtained by other authors [90, 99–101].

Calculations of the equilibrium Ar – He plasma composition within the temperature range of $300 \text{ K} \le T \le 25000 \text{ K}$ show that, at pressures close to atmospheric, the plasma considered is transparent even for the lowest-frequency CO_2-laser radiation, since this case involves meeting the $\omega > \omega_{pe}$ condition, where

$$\omega_{pe} = \sqrt{n_e e^2 / m_e \varepsilon^0}$$

is the electron plasma frequency. Therefore, to determine $\varepsilon_\omega (T, p) = \varepsilon_\omega' (T, p) + i\varepsilon_\omega'' (T, p)$, we will use an expression for the complex permittivity of the multicomponent "cold" plasma [102], which, at $\omega > \omega_{pe}$, can be written as follows:

$$\varepsilon_\omega = 1 - \frac{\omega_{pe}^2}{\omega (\omega + iv_e)} , \qquad (2.54)$$

where v_e is the total frequency of collisions of electrons with heavy particles. Taking into account that $\omega \gg v_e$ in the case considered, the real and imaginary parts of the complex permittivity of the plasma will be expressed through relationships [103]:

$$\varepsilon_\omega' = 1 - \frac{\omega_{pe}^2}{\omega^2} ; \qquad (2.55)$$

$$\varepsilon_\omega'' = \frac{ck_\omega \sqrt{\varepsilon_\omega'}}{\omega} , \qquad (2.56)$$

where c is the velocity of light.

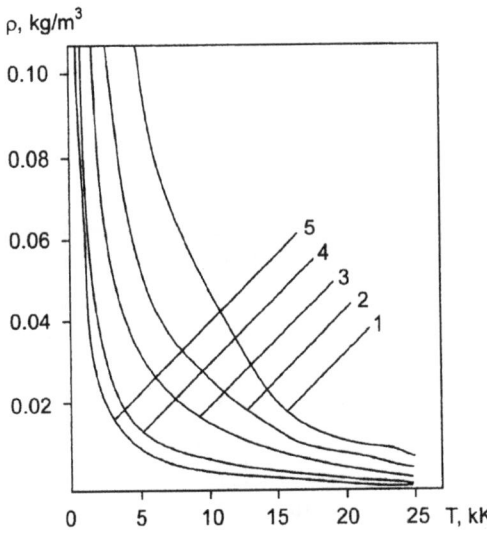

Figure 2.2 Variation of the mass density of Ar – He plasma with temperature at atmospheric pressure: 1, 100 % Ar; 2, 50 % Ar + 50 % He; 3, 25 % Ar + 75 % He; 4, 5 % Ar + 95 % He; 5, 100 % He.

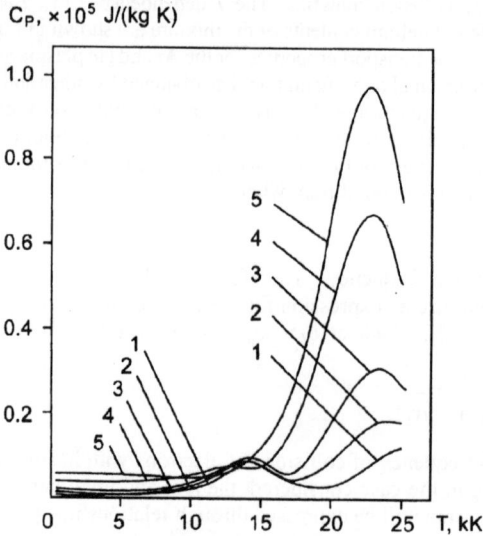

Figure 2.3 The specific heat capacity at constant pressure of Ar – He plasma at 1 atm as a function of temperature (parameters and designations are the same as in Figure 2.2).

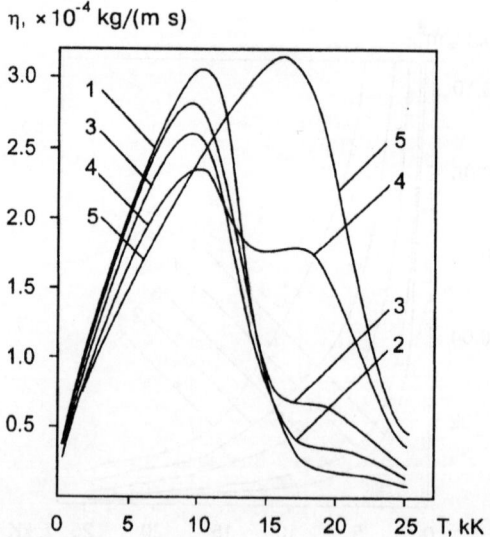

Figure 2.4 The dynamic viscosity coefficient of Ar – He plasma at 1 atm as a function of temperature (parameters and designations are the same as in Figure 2.2).

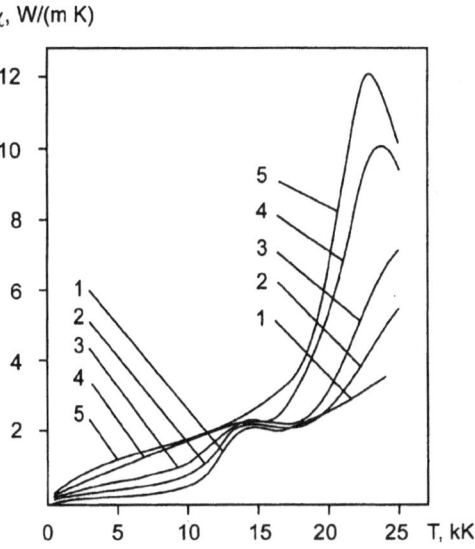

Figure 2.5 The thermal conductivity coefficient of Ar – He plasma at 1 atm as a function of temperature (parameters and designations are the same as in Figure 2.2).

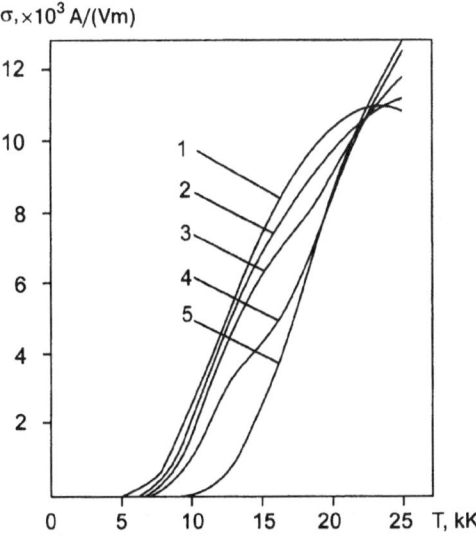

Figure 2.6 Variation of the specific electrical conductivity of Ar – He plasma with temperature at 1 atm (parameters and designations are the same as in Figure 2.2).

Calculation of the laser radiation absorptivity $\kappa_\omega(T, p)$ of the equilibrium Ar–He plasma was performed by the Biberman-Norman formulae [104], which allow both for the free-free transitions (inverse Bremsstrahlung absorption) and the bound-free ones (photoionisation from the excited levels). In the infrared region of the spectrum, the expression for κ_ω corrected for stimulated emission of the plasma has the form [104, 105]:

$$\kappa_\omega = \frac{e^6 n_e \left[1 - \exp\left(- \frac{\hbar\omega}{k_B T} \right) \right]}{48 c \hbar \sqrt{6\pi^7 m_e^3 k_B T}} \sum_{\alpha = 2, 3, 5} \xi_\alpha Z_\alpha^2 n_\alpha \exp\left(- \frac{\Delta U_\alpha}{k_B T} \right) \times$$

$$\times \left\{ \exp\left[\frac{\hbar(\omega + \Delta\omega_\alpha)}{k_B T} \right] - 1 + g \right\} \tag{2.57}$$

where: $\xi_\alpha(\omega)$ is a certain frequency function [106]; $\Delta\omega_\alpha$ is the optical shift of the continuous spectrum boundaries for atoms and singly charged ions of Ar ($\alpha = 2, 3$) and atoms of He ($\alpha = 5$) [107];

$$g = \frac{\sqrt{3}}{\pi} \ln\left(\frac{16\pi\varepsilon^0 k_B T}{e^2 n_e^{1/3}} \right)$$

is the Gaunt factor.

It follows from formula (2.57), that the laser radiation absorptivity of the plasma largely depends on the radiation frequency. For equilibrium atmospheric pressure argon plasma absorption of the CO_2-laser radiation ($\lambda = 10.6$ microns) is approximately two orders higher

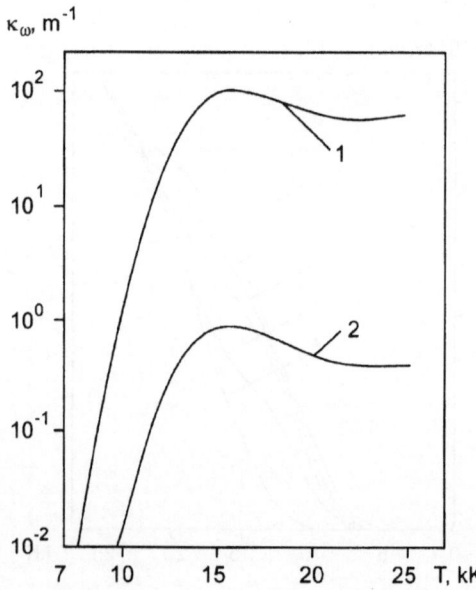

Figure 2.7 Comparison of the absorptivity of argon plasma at atmospheric pressure for different kinds of laser radiation: 1, CO_2-laser ($\lambda = 10.6$ μm); 2, Nd:YAG-laser ($\lambda = 1.06$ μm).

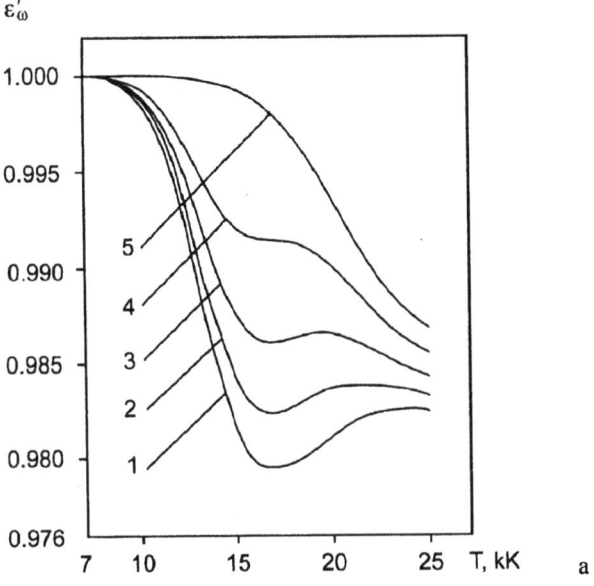

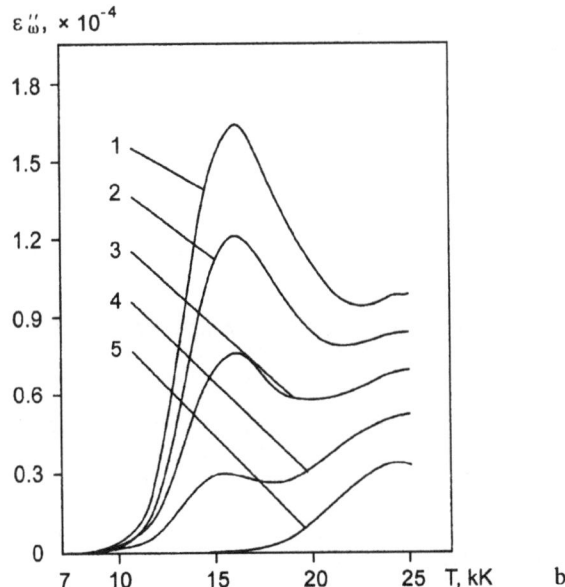

Figure 2.8 Real part (a) and imaginary part (b) of the complex permittivity of Ar – He plasma at 1 atm for CO_2-laser radiation as a function of temperature (parameters and designations are the same as in Figure 2.2).

Ψ, $\times 10^{10}$ W/m^3

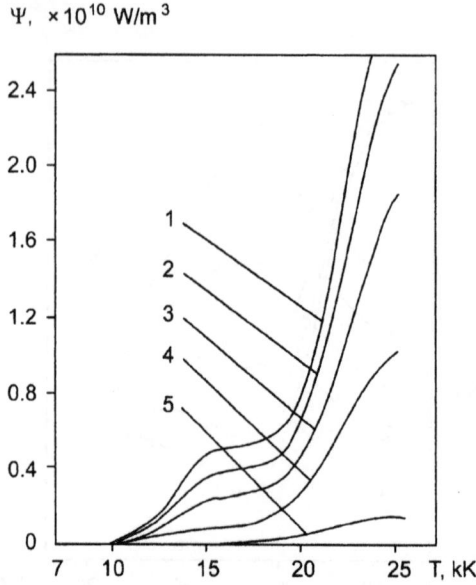

Figure 2.9 Variation of the radiative losses of Ar – He plasma with temperature at 1 atm (parameters and designations are the same as in Figure 2.2).

(Figure 2.7), than that of the Nd:YAG-laser ($\lambda = 1.06$ micron). Besides, the optical properties κ_ω and ε_ω greatly depend on temperature, pressure and composition of the plasma. As an example, Figure 2.8 shows the temperature dependences of $\varepsilon_\omega{}'$ and $\varepsilon_\omega{}''$ calculated from formulae (2.55) to (2.57) at the CO_2-laser radiation frequency for the Ar – He plasma at atmospheric pressure and with different helium contents of the mixture.

Irradiation of the plasma under consideration is made up of continuous (recombination–slowing down continuum) and bright-line emission spectra, the main role being played by the recombination radiation [105]. Therefore, to determine the total radiation energy losses, it is possible to use the results of paper [104], representing the expression for the spectrum-integral radiative power of the Ar – He plasma in the following form [108]:

$$\psi = \frac{e^6 \sqrt{k_B T}\, n_e}{6c^3 \hbar \sqrt{6\pi^5 m_e^3}\, \varepsilon^{0^3}} \sum_{\alpha = 2, 3, 5} Z_\alpha^2 n_\alpha \exp\left(\frac{\hbar \omega_\alpha - \Delta U_\alpha}{k_B T} \right),$$

(2.58)

where ω_α are the experimental effective frequencies of the integral levels [109] for atoms and singly charged ions of Ar ($\alpha = 2, 3$), as well as atoms of He ($\alpha = 5$). Results of calculations of $\psi(T)$ for the equilibrium atmospheric pressure Ar – He plasma are shown in Figure 2.9. It should be noted in conclusion, that the results obtained are in satisfactory agreement with the data available in the literature on arc plasma radiation, despite the contradictory character of the latter (see, e.g. [72, 75, 109]).

2.3 Analysis of the Results of Laser-Arc Discharge Simulation

The present section deals with numerical modelling of the stationary combined discharge induced by a CW CO_2-laser beam affecting the DC arc burning in the axial flow of inert gases (Ar, He or their mixtures). The effects are studied of power and focusing of the laser radiation, arc current, flow rate and composition of the plasma gas, as well as the geometrical dimensions of the channel, on the distributed and integral characteristics of the combined discharge plasma and the laser beam propagating in it.

In the numerical realisation of the laser-arc discharge mathematical model described in Section 2.1, the arc current was varied within the range of 5 A $\leq I \leq$ 250 A, and the laser radiation power within the range of $0 \leq Q^0 \leq$ 1.5 kW. The distance F of the initial beam focus plane from the initial section of the calculated region was selected as 5, 10, 15 and 20 mm, while the minimum beam radius r_F was changed from 0.2 to 0.5 mm. The relative accuracy achieved in the calculations was no worse than 5 %.

2.3.1 Laser beam effect on the arc plasma inside the channel

Consider the effect of a focused CO_2-laser radiation beam on the plasma of the argon arc burning inside a cylindrical channel with axial gas flow (see Figure 2.1) under the following conditions. The channel length L_C is 4 cm, its radius R_C is 2.5 mm, pressure at the channel outlet is atmospheric, total mass flow rate of argon through the channel is 0.05 g/s $\leq G \leq$ 0.1 g/s, the cathode nozzle radius R_1 is 1.0 mm, argon flow rate through the internal nozzle cavity is 0.01 g/s $\leq G_1 \leq$ 0.025 g/s, and the temperature of the cooled wall of the channel and the nozzle is $T_C = T_1 = 300$ K. The length L of the discharge calculated domain (see Figure 2.1) is equal to the length of the channel and, as shown by calculations, is sufficient to reveal peculiarities of the laser-arc discharge plasma in all three regions typical of this discharge at the selected values of L_C and other parameters.

Figures 2.10–2.20 show the results of numerical study of distributed and integral characteristics of the combined discharge plasma inside the channel. In particular, Figure 2.10 shows the radial distributions of the sources for heating the plasma by the current (j^2/σ) and the absorbed laser radiation ($\kappa_\omega S$) within region II (region $Z_I \leq z \leq Z_{II}$ in Figure 2.1), as well as the distribution of the energy losses due to plasma irradiation (ψ). The given calculated dependences show the higher efficiency of laser beam heating of the plasma, as compared to its Joule heating, the energy of the beam, because of its small transverse sizes, being introduced locally into the plasma (compare Figures 2.10a and b). Note, that not only the non-proportional increase in energy put into the discharge by the laser radiation, but also the marked redistribution of the sources j^2/σ and ψ (see Figures 2.10a and c) associated with a change in the plasma temperature, take place with an increase in the laser radiation power.

Spatial distributions of temperature of the arc plasma inside the channel affected by the laser radiation are given in Figure 2.11, as compared to the corresponding distributions for conventional argon plasma arc. As follows from Figure 2.11a, action of the laser beam on the arc column plasma leads to an increase in temperature of its central region, this being caused by local heating of the plasma by the laser beam. Starting from a certain section $z = Z_I$, determining the boundary between regions I and II of the discharge under investigation, the temperature $T_0(z)$ at the channel axis drastically increases (see Figure 2.11b), reaching the value that is much in excess of the temperature of the conventional arc plasma in the corresponding sections of the channel. Then, coming over the maximum, with increase in z, T_0 starts gradually decreasing, approach-

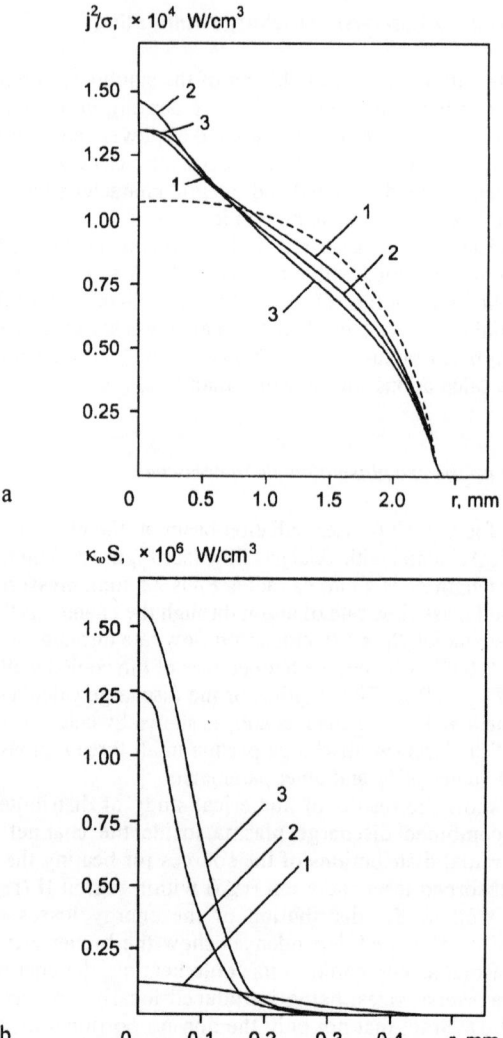

Figure 2.10 Radial distributions of the Joule heating (a), the laser radiation absorption (b) and the radiative losses (c) in the combined discharge plasma ($I = 100$ A; $R_C = 2.5$ mm, $R_1 = 1.0$ mm; $G = 0.1$ g/s, $G_1 = 0.01$ g/s; $r_F = 0.2$ mm, $F = 10$ mm) at $z = 10$ mm for different laser beam powers: 1, $Q^0 = 0.5$ kW; 2, 1.0 kW; 3, 1.5 kW; dashed curves, $Q^0 = 0$.

ing its asymptotical value*. A drop of the axial temperature taking place in the conversion region III (region $z > Z_{II}$ in Figure 2.1) is related, as noted above, to a decrease in the laser beam heating of the plasma due to attenuation of the beam when propagating in an

---

* Radial distributions of the plasma characteristics in asymptotical region of the discharge were calculated on the basis of numerical solution of one-dimensional equations for the totally developed flow of the arc plasma inside the channel (see, e.g. [75]).

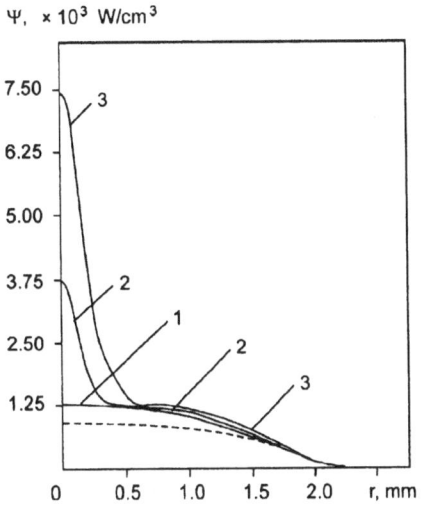

Ψ, $\times 10^3$ W/cm^3

Figure 2.10 (*continued*).

absorbing medium. Therefore, a relatively high-temperature region localised near the discharge axis forms in the initial region of the argon arc burning in the channel under the effect of the focused CO_2-laser radiation, and the maximum temperature of the plasma achieved in the laser-arc discharge increases with growth of the laser beam power (see Figure 2.11b).

Growth of the degree of ionisation and, hence, of the plasma electrical conductivity (see Figure 2.6) caused by an elevation of the temperature should lead, according to integral relationship (2.5), to a decrease in the field intensity necessary to maintain the preset current. Distributions of the electric field intensity along the channel length shown in Figure 2.12 prove in full this assumption. It should be noted that the relative decrease in the E value due to additional heating of the arc plasma by the laser beam, is aggravated with growth of its power.

The local increase in plasma conductivity and the drop of the electric field intensity in the discharge occurring under the effect of the laser beam on the arc, result in significant redistribution of the current flowing in it (Figure 2.13). Under the conditions studied, the increase in electrical conductivity near the arc axis is a more significant factor than the decrease in the field intensity in the plasma constant over the discharge section. Therefore, the axial component of current density calculated by using relationship (2.4) grows in the centre under the effect of the laser radiation and somewhat decreases near the channel walls (see Figure 2.13a). Distribution of current density at the discharge axis $j_0(z)$ along the channel corresponds to behaviour of the axial plasma temperature (compare Figures 2.13b and 2.11b), the difference being due to decrease in the field intensity. The maximum j_0 shifts to the initial section and becomes more dramatic, while the maximum achievable value of the current density does not significantly increase with the growth of the laser beam power (see Figure 2.13b).

Calculations of spatial distributions of T and j within the discharge investigated for different values of the r_F and F parameters show, that a reduction of the maximum radiation intensity $S_{max}^0 = 2Q^0/(\pi r_F^2)$ in the initial laser beam due to an increase in r_F with constant Q^0, leads to a marked reduction of the values of the maximum temperature and current density. A change (within the selected limits) in the position of the initial beam neck, i.e. in

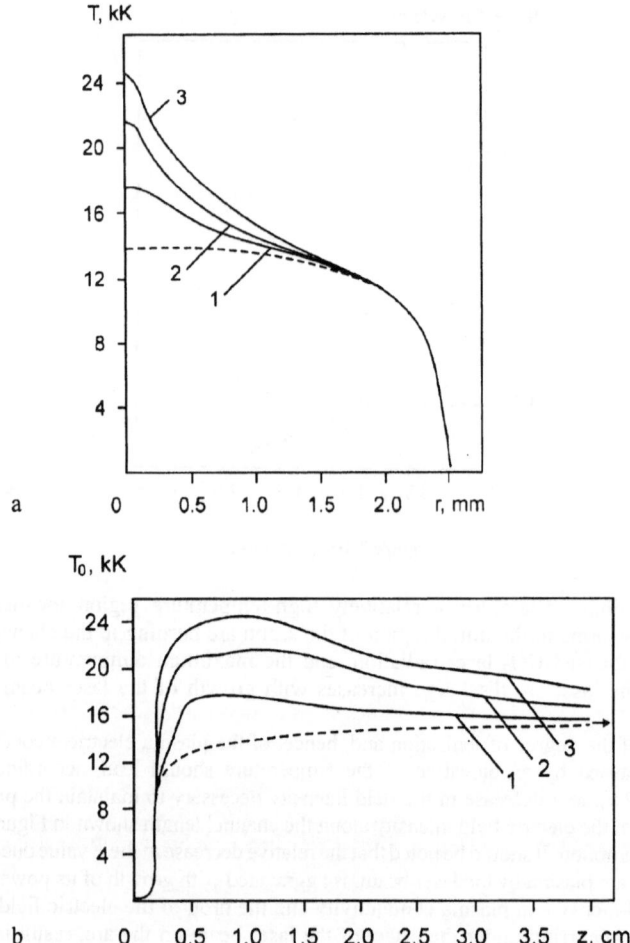

Figure 2.11 Radial profiles of the plasma temperature at $z = 10$ mm (a) and distributions of the temperature on the discharge axis along the channel (b) for different Q^0 (parameters and designations are the same as in Figure 2.10). The arrows in Figures 2.11b, 2.12, 2.13b, 2.14c, 2.15c indicate the values of the corresponding characteristics in asymptotical region.

the F distance, with fixed values of r_F and Q^0, affects to a much lower degree the spatial distributions of the plasma temperature and, thus, current density in the discharge. This can be explained by the fact that for the slightly focused beams under investigation ($\vartheta_F = \lambda/(\pi r_F) < 0.02$) the radiation intensity near the focus is a rather slowly varying function of z.

A change in the thermal conditions of the arc discharge burning under the effect of focused laser radiation leads not only to the above-mentioned variation of electric properties of the arc column, but also to a considerable redistribution of the hydrodynamic characteristics of the plasma flow (Figures 2.14, 2.15). One of the reasons for it is a

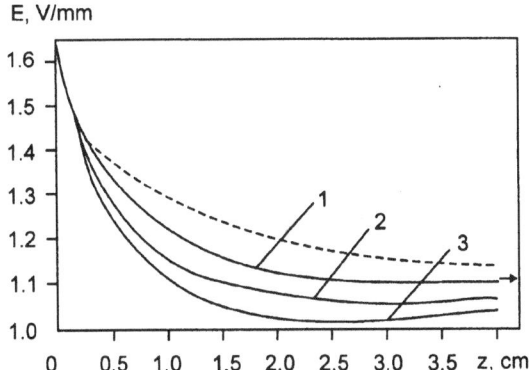

Figure 2.12 Longitudinal distributions of the electric field in the discharge for different Q^0 (parameters and designations are the same as in Figure 2.10).

drastic decrease in the argon viscosity coefficient at $T > 10000$ K (see Figure 2.4), which, according to equation (2.2), causes a reduction of the viscous forces that slow down the motion of the plasma in the centre and accelerate the gas on the arc periphery. As a result, the axial component u of the plasma velocity grows, its growth with increase in the laser beam power being most significant near the channel axis (see Figure 2.14a). Besides, redistribution of the current density occurring in the discharge under the effect of the laser radiation promotes a growth of the role played by the electromagnetic forces in the formation of the plasma flow in the near-axis zone of the arc. Here, there occurs not only a variation in the values, but also a variation in the directions of the Lorentz forces that slow down the plasma in the initial sections of the discharge and accelerate it in the region of a decrease in the j_0 value. All this leads to the situation that the longitudinal velocity component $u_0(z)$ at the arc axis, unlike the axial temperature, grows with increase in z more smoothly (compare Figures 2.11b and 2.14c) and reaches its asymptotical value at a much greater distance along the flow.

Along with increase in the axial velocity component, increase in the temperature of the plasma results in a decrease in its density (see Figure 2.2) and, hence, redistribution of the mass velocity ρu (see Figure 2.15a) and the gas dynamic pressure in the flow $\rho u^2/2$ (see Figure 2.15b). In particular, a deepening of the "cavity" ρu with the growth of the laser beam power is observed at the discharge axis (see Figure 2.15a). The gas necessary to fill it is introduced from the peripheral regions of the column, resulting in an increase in the absolute value of the radial velocity v, which is negative (see Figure 2.14b). Unlike the mass velocity, the gas dynamic pressure, being a root mean square function of u, decreases in the near-axis regions less significantly and only at the initial sections of the discharge (see Figure 2.15c). The value of $\rho_0 u_0^2/2$ (at the discharge axis) grows with distance from the channel inlet, reaching a maximum, and, then, at a rather high z, tends to the asymptotical value characteristic of the conventional arc inside the channel.

Evolution of the radial distributions of the plasma characteristics for the discharge under consideration along the channel is shown in Figure 2.16. A distinctive feature of region I (region $z < Z_1$ in Figure 2.1) is the presence of a central flow of cold gas and a circumferential flow of the arc plasma coaxial to it (curves 1 in Figures 2.16a–c), the intensive flow out of the gas from the central regions to the channel periphery (curve 1 in Figure 2.16d) and its entrapment into the arc being observed here due to essential heterogeneity of distribution of

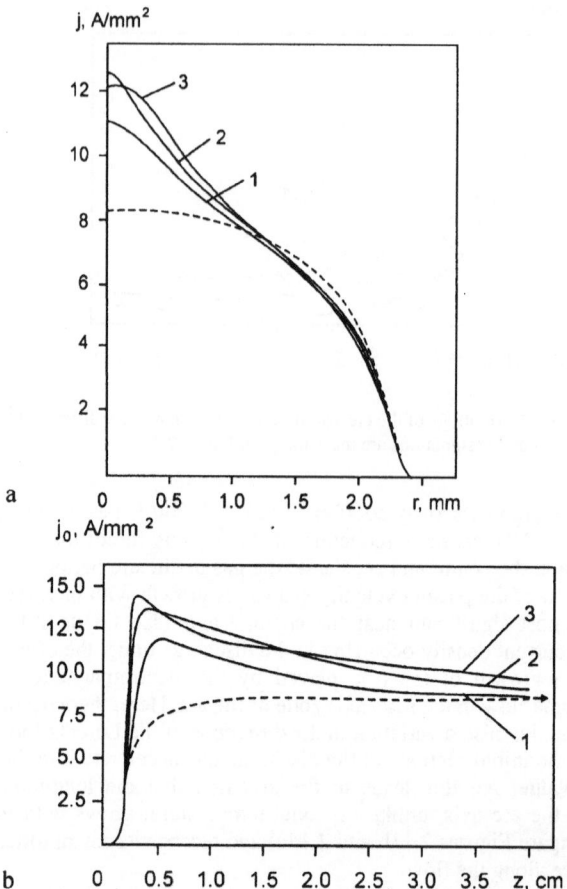

Figure 2.13 Radial profiles of the axial current density at $z = 10$ mm (a) and distributions of the current density on the discharge axis along the channel (b) for different Q^0 (parameters and designations are the same as in Figure 2.10).

the mass velocity along the radius at the selected ratio of G_1 and G_2 (curve 1 in Figure 2.16e). At a distance from the inlet section the effect of the viscous forces leads to an increase in the near-axis velocity of the gas flow (compare curves 1 and 2 in Figure 2.16c). However, due to decrease in its density with growth of the temperature at the axis (curve 2 in Figure 2.16a) the maximum of the mass velocity shifts to the channel wall (curve 2 in Figure 2.16e), resulting in an alternation of the sign of the radial velocity (curve 2 in Figure 2.16d). Heating up of the cold gas near the axis causes a reduction of the radius of the non-conducting zone and, at some distance Z_1 from the channel inlet, the arc collapses, filling in the entire channel section (curves 2 in Figures 2.16a, b).

At $z = Z_1$, the focused laser radiation propagating in the cold gas practically without any absorption enters the plasma. This implies an end of the undisturbed region of the arc column and a beginning of the laser-arc interaction region. A change in the arc column heat balance described above is observed here due to additional local heating of the plasma by the laser

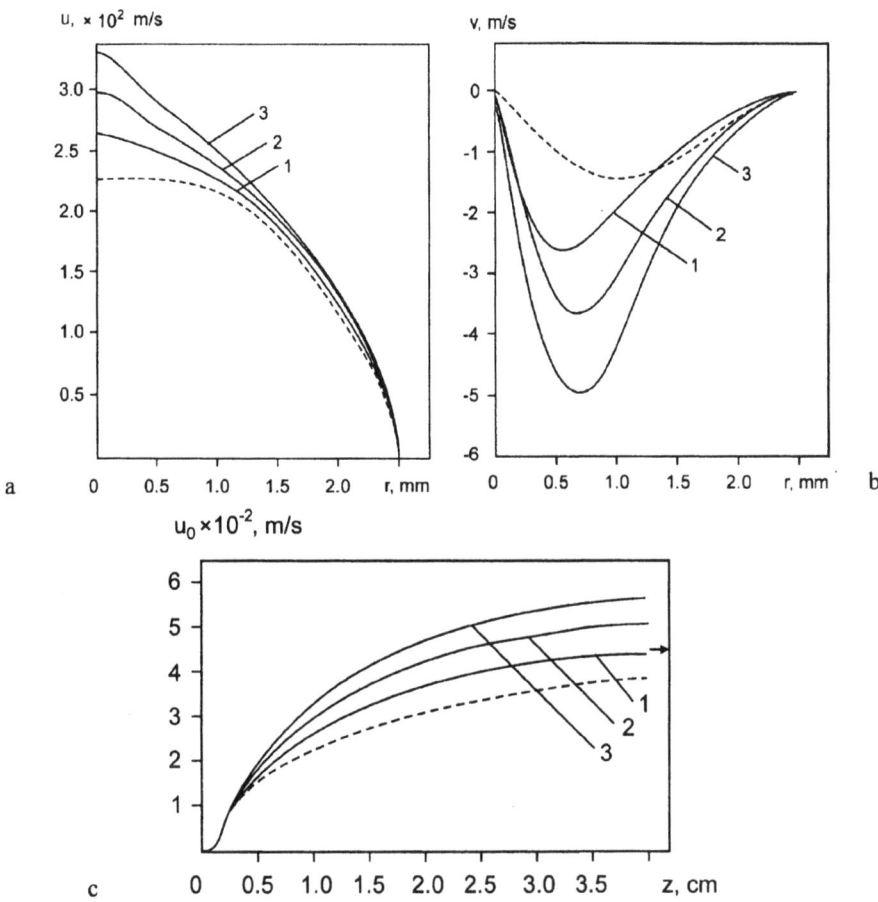

Figure 2.14 Radial profiles of axial (a) and radial (b) components of the plasma velocity at $z = 10$ mm and distributions of the velocity on the discharge axis along the channel (c) for different Q^0 (parameters and designations are the same as in Figure 2.10).

beam. As a result, distributions of the plasma characteristics over the channel section (curves 3 in Figures 2.16a–e) take a form typical of laser-arc discharge (see, e.g., Figures 2.11a, 2.13a, 2.14a, 2.15a).

The intensity of the laser radiation diminishes due to the processes of absorption and refraction with distance from the point where the beam enters into the plasma. This leads to a gradual decrease in its role in the energy balance of the discharge. Thermal disturbance induced in the plasma by the laser beam starts relaxing (curve 4, Figure 2.16a), current density at the discharge axis drops (curve 4, Figure 2.16b), indicating the beginning of the conversion region, wherein the most significant variations in the gas dynamic characteristics of the plasma flow take place. In particular, the axial velocity grows up to values in excess of the asymptotical ones (curve 4, Figure 2.16c), passes through the maximum and, then, starts decreasing, approaching its asymptotical distribution (dashed curve in Figure 2.16c). The mass velocity ρu near the channel axis, unlike u, grows monotonically, leading to smoothing out of its profile (curve 4, Figure

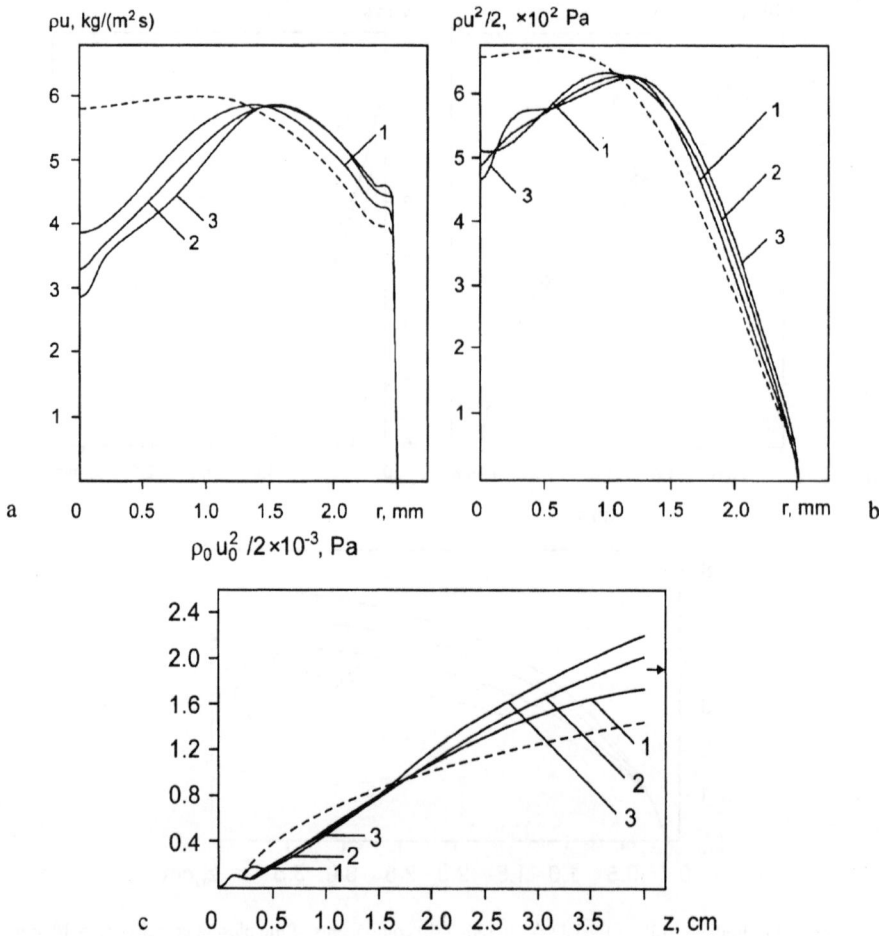

Figure 2.15 Radial profiles of axial component of the mass velocity (a) and the gas dynamic pressure of the plasma flow (b) at $z = 10$ mm and distributions of the gas dynamic pressure on the discharge axis along the channel (c) for different Q^0 (parameters and designations are the same as in Figure 2.10).

2.16e) and, as a result, to a decrease in the absolute values of negative radial plasma velocities (curve 4, Figure 2.16d).

The length of region I, i.e. the distance from the inlet section to the beginning of the laser-arc interaction region*, depends on the arc current, the laser beam parameters and, to a great degree (see Figures 2.17a–c), on the plasma gas flow pattern at the inlet to the channel. Thus, for example, with increase in the cold gas flow rate G_1 through the nozzle opening, the arc column is forced to the channel wall at the initial sections. This results in some growth

---

* A section where the plasma temperature on the channel axis is higher than 11000 K, where the marked laser radiation absorption is observed, is assumed to be the beginning of the region of interaction of the CO_2-laser radiation and the argon plasma under a pressure close to atmospheric.

of the field intensity (compare curves 1 and 2 in Figure 2.17b) and in the collapse of the plasma taking place much farther along the flow (see Figures 2.17a and c). Therefore, by varying the value of G_1 it is possible to vary, within wide enough ranges, the position of introduction of the laser beam into the plasma, i.e. the beginning of the laser-arc interaction region. Variation in the total plasma gas flow rate G through the channel does affect, but to a lesser degree, the length of the arc column region, which does not interact with the laser radiation (see Figures 2.17a and c). It should be noted that a slower growth of the plasma temperature at the axis (compare dashed and solid curves in Figure 2.17a) and a smoother drop of the field intensity (see Figure 2.17b) are observed with increase in G, the maximum temperature achieved in the discharge reducing to some extent. And, by contrast, the axial component of the plasma velocity starts growing more rapidly (see Figure 2.17c), and its maximum value increases almost proportionally with the total mass flow rate of the plasma gas through the channel.

The length of the laser-arc interaction region (region II) is determined by the degree of laser radiation absorption of the plasma. Absorptivity $\bar{\kappa}_\omega (z)$ averaged over the laser beam section, calculated with the help of relationship:

$$\bar{\kappa}_\omega = \frac{2\pi}{Q} \int_0^{R_C} \kappa_\omega \, Sr \, dr \,, \tag{2.59}$$

greatly depends on the spatial distribution of the plasma temperature in the discharge. Distribution of the temperature, as well as of other characteristics of the combined discharge plasma inside the channel depends not only on the laser radiation power, as shown above, but also on the value of the arc current*, this being illustrated in Figure 2.18. Also, this Figure shows the parameter $\gamma (z)$ varying along the channel length, that characterises the ratio of the energy, released within a certain section of the discharge due to the electric current, to the total energy supplied to the plasma within the same section:

$$\gamma = \frac{EI}{EI + \bar{\kappa}_\omega Q} \,. \tag{2.60}$$

A decrease in current, other conditions being equal, leads to an elongation of region 1 and, accordingly, to an increase in distance of the beginning of the laser-arc interaction region from the inlet section (see Figure 2.18a). Of special note is the fact that at very low currents ($I < 10$ A) and low powers of the laser radiation ($Q^0 < 0.7$ kW) the formation of the combined discharge in argon is not possible at all, since the maximum temperature of the argon arc plasma in the channel at $I/D < 2$ A/mm does not exceed 10000 K and the absorption of the CO_2-laser radiation is negligibly low (see Figure 2.7). However, it is in the case of low currents (10 A $< I < 25$ A), providing the power of the laser beam is sufficient for the formation of the laser-arc discharge, that the most drastic relative variations of the axial temperature, the field intensity and the axial plasma velocity take place (compare e.g., dashed and solid curves 1 in Figures 2.18a, b and d). Because the thermal disturbance propagation velocity is finite, under the effect of local heating of the arc plasma by the laser radiation the growth of T_0 along z occurs faster than the decrease in the value of E determined by the distribution of the temperature over the entire channel section. As a result, the current density at the axis first grows, as compared to the corresponding values for the conventional arc (curves 1 in Figure 2.18c), reaches a maximum and, then, drastically drops down to the values located below the j_0 level in the asymptotical region of the channel. Thus, at the low arc

---

* The most convenient criterion for the current value for the discharge inside the channel is the ratio I/D ($D = 2R_C$—channel diameter) determining the degree of arc constriction [73].

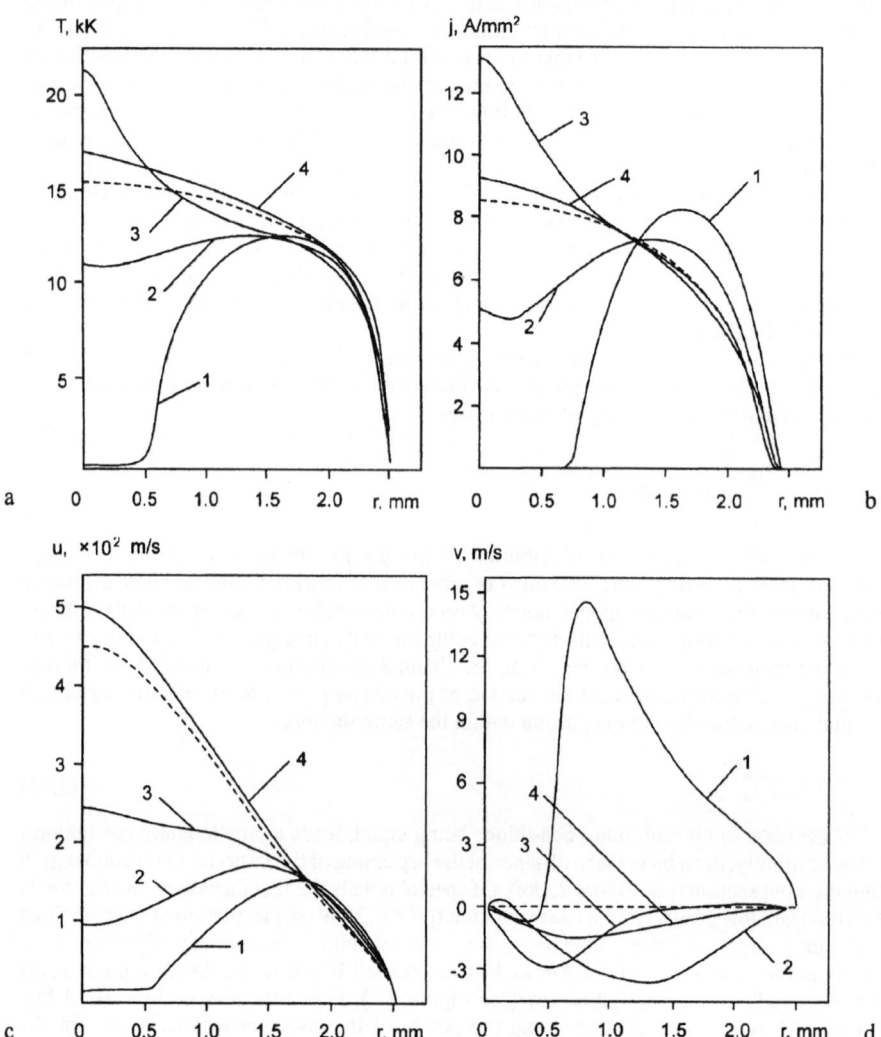

Figure 2.16 Radial distributions of the plasma temperature (a), the current density (b), axial (c) and radial (d) components of the plasma velocity and axial component of the mass velocity (e) for different cross-sections of the channel ($G_1 = 0.025$ g/s; $Q^0 = 1.0$ kW; other parameters are the same as in Figure 2.10): 1, $z = 1$ mm; 2, 5 mm; 3, 10 mm; 4, 40 mm; dashed curves, the distributions of the corresponding characteristics in asymptotical region.

currents within the initial sections of the laser-arc interaction region, a region forms with a higher current density at the axis. Note that the presence of this region cannot be explained within the framework of a one-dimensional model [65, 66]. Significant electromagnetic forces originating in the region of drastic drop of the j_0 value, which act in the positive direction of the OZ axis, and a decrease in the plasma viscosity as a result of the elevation

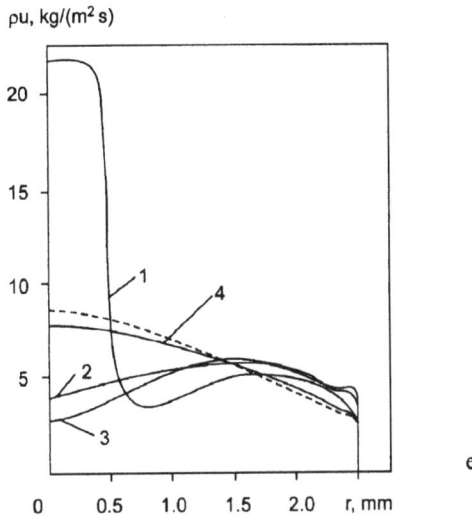

Figure 2.16 (*continued*).

of the plasma temperature lead to a sudden increase in the axial components of the flow velocity, as compared to a conventional arc discharge (curves 1 in Figure 2.18d).

The higher the arc current, the sooner the collapse of the annular plasma flow entering the channel occurs (curves 2 and 3 in Figure 2.18a), i.e. the laser-arc interaction region approaches the inlet section. Besides, a decrease in field intensity in the discharge with growth of the current under the effect of the laser radiation becomes a less significant factor than an elevation of the temperature of the plasma in the central regions. As a result, the current density near the channel axis (curves 2 and 3 in Figure 2.18c) along the entire region II remains higher than the current density in the conventional arc. The maximum of the j_0 value also shifts to the initial section with the growth of the total current, and somewhat smoothes out (see Figure 2.18c). The change in the u_0 value under the effect of the laser radiation becomes less noticeable against the general growth of the axial component of the arc plasma velocity with the growth of I (see Figure 2.18d).

The above peculiarities in behaviour of the axial distributions of the combined discharge plasma characteristics observed with a variation in current can be compared with different characters of the $\gamma(z)$ dependences in the cases of low and high currents (see Figure 2.18e). Taking into account that the γ value in each of the channel sections characterises the relative contributions of the arc current and the laser radiation to the plasma energy balance, we can determine the region of laser-arc interaction (region II) such that $\gamma < 0.9$. It will mean that in any section of the said region the laser beam will introduce into the plasma not less than 10 % of the energy input. It can be seen from the calculated dependences shown in Figure 2.18e, that the position of the beginning of region II and its length, greatly depend on the arc current. The most significant variations along the channel length of the parameter studied are observed for low values of I. A sudden decrease in γ occurring at small currents in the initial section of the laser-arc interaction implies that the discharge maintained mostly by the electric current ($\gamma > 0.5$) transforms here practically in a jump-like manner into the other type of discharge existing mostly due to absorption of the laser radiation in the plasma ($\gamma < 0.5$). Thus, for example, at $I = 10$ A (curve 1 in Figure 2.18e) there is a discharge region, where $\gamma < 0.1$, i.e. more than 90 % of energy is supplied to

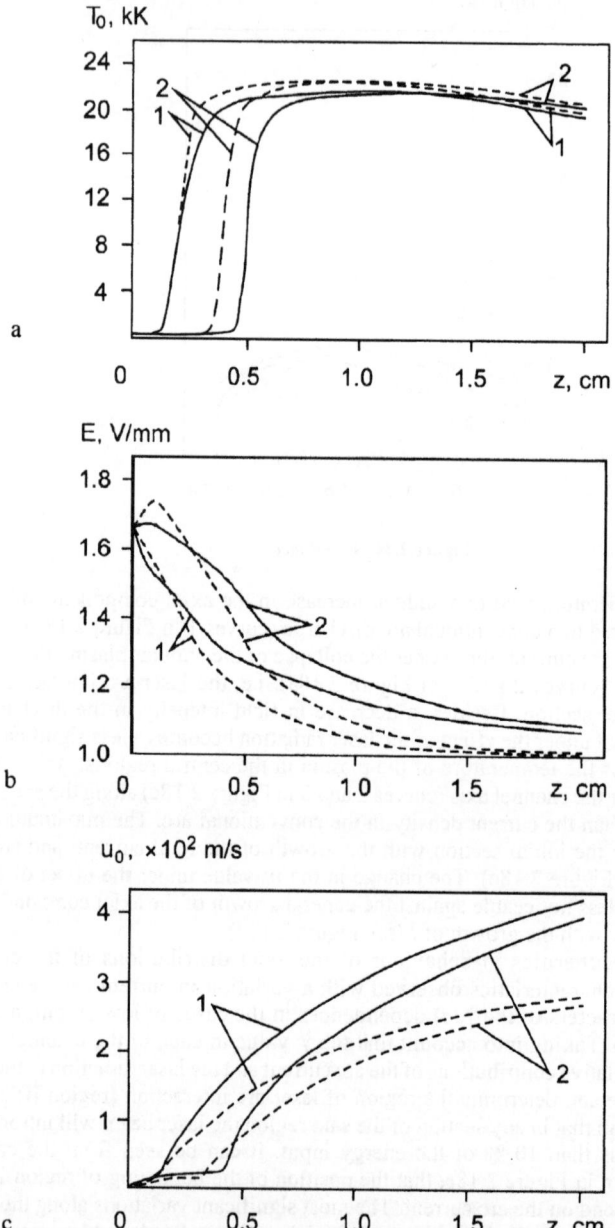

Figure 2.17 Longitudinal distributions of the plasma temperature on the discharge axis (a), the electric field (b) and the axial velocity of the plasma (c) for different plasma gas flow rates ($Q^0 = 1.0$ kW): 1, $G_1 = 0.01$ g/s; 2, 0.025 g/s; solid curves, $G = 0.1$ g/s; dashed curves, 0.05 g/s (other parameters are the same as in Figure 2.10).

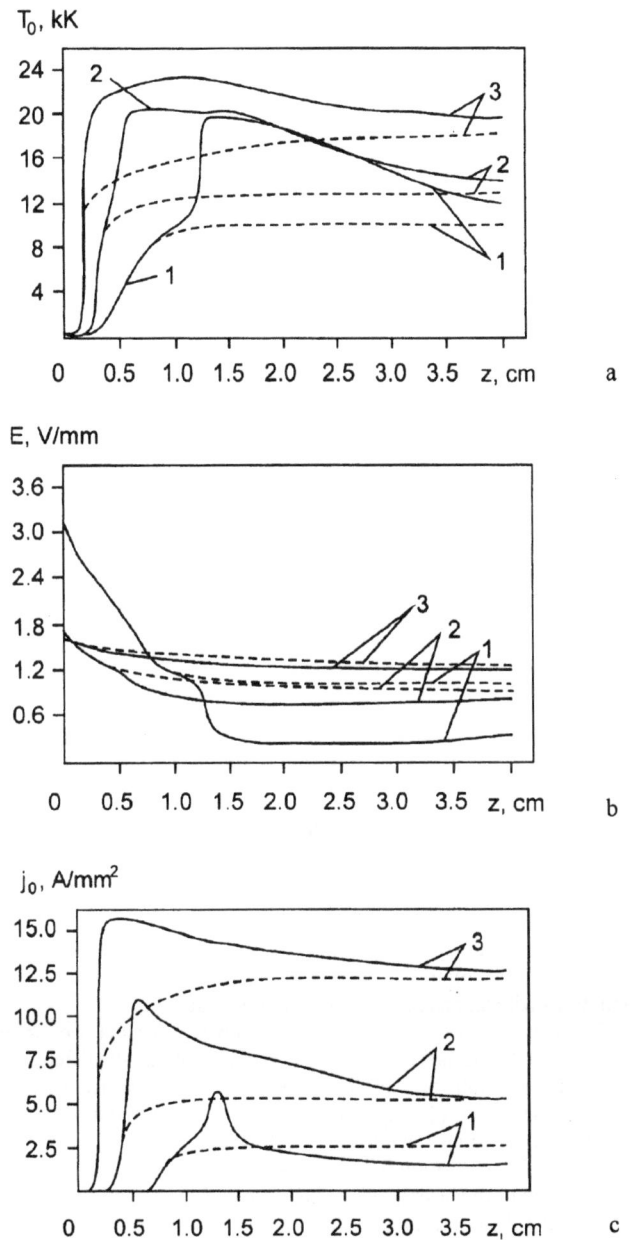

Figure 2.18 Longitudinal distributions of the plasma temperature on the discharge axis (a), the electric field (b), the axial current density (c), the axial velocity of the plasma (d) and the parameter γ (e) for different arc currents: 1, I = 10 A; 2, 50 A; 3, 150 A; solid curves, Q^0 = 1.0 kW; dashed curves, Q^0 = 0 (other parameters are the same as in Figure 2.10).

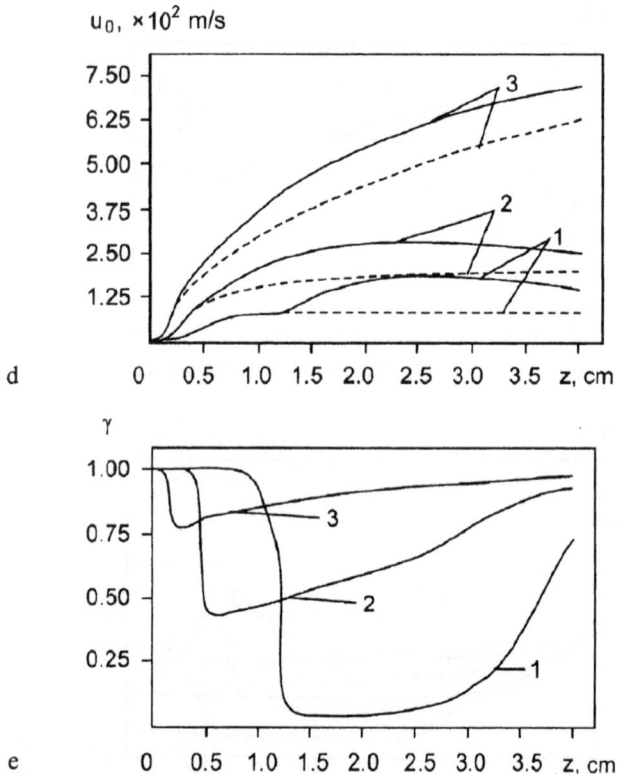

u_0, $\times 10^2$ m/s

d

γ

e

Figure 2.18 (*continued*).

the plasma by the laser beam. Since with an increase in distance passed by the beam in the laser-arc discharge plasma, the share of the laser radiation in the plasma energy balance decreases due to attenuation of the beam, the parameter γ, upon passing through its minimum value, starts growing, approaching its asymptotical value of γ = 1. The rate of its growth also depends on the arc current and drops with an increase in I (see Figure 2.18e).

Note the fact, that at small currents in the region where plasma is maintained mostly by laser radiation (γ < 0.1) the laser-arc discharge must not be identified in the case under consideration with continuous optical discharge in the gas flow, because, with the selected parameters of the laser beam (Q^0, ϑ_F) and values of the inflowing gas rate, the stationary optical discharge in argon does not exist [78]. At the same time, even the presence of a low arc current sufficient for realisation of the laser-arc discharge makes it possible to considerably expand the region of the laser plasma existence and use a combined discharge to induce the plasma flow under the conditions, where the laser plasmatron mode can not be realised.

Local changes in the plasma energy balance occurring under the effect of laser radiation become less sudden with growth of current (curves 2 and 3 in Figure 2.18e). It is associated with the fact that, with an increase in I, the gap between the absolute values of the power input per unit length of the arc and the combined discharge decreases. As shown by calculations, at I = 10 A about 100 W/cm are introduced into the section z = 20 mm of the arc column, whereas at Q^0 = 1.0 kW about 480 W/cm are put into the same section of the combined discharge. If I = 150 A, 2040 and 2080 W/cm, respectively, are introduced. Therefore, the efficiency of the additional laser heating of the electric arc plasma inside the channel decreases with growth of current, i.e. with an increase in a degree of the arc column constriction.

Along with the parameter considered above, to estimate the relative role of different mechanisms of plasma heating in the total energy balance of the laser-arc discharge column, the following value can be used:

$$\gamma_L = \frac{U_L I}{U_L I + Q_L}. \tag{2.61}$$

Here, U_L is the potential drop in the discharge region having length L; Q_L is the laser radiation power absorbed in the same region, which can be determined by relationships:

$$U_L = \int_0^L E \, dz; \tag{2.62}$$

$$Q_L = \int_0^L \bar{\kappa}_\omega Q \, dz. \tag{2.63}$$

Figure 2.19 shows dependences of the γ_L value on the laser beam power for the discharge region 2 cm long at different arc currents, while Figure 2.20 presents dependences of the voltage drop U_L in the region under consideration on the arc current at different Q^0. As follows from the calculated data given in Figure 2.19, an increase in the power of the initial beam leads to a growth of its role in the total energy balance of the initial region of the discharge, this being associated with an increase in the laser radiation power Q_L absorbed by the plasma and with a simultaneous decrease in the electric power $U_L I$ due to a drop of voltage with a growth of Q^0 in the same region (see Figure 2.20). In the case when I = 10 A, against a gradual decrease in γ_L with an increase in the laser radiation power, there occurs a drastic drop of the said value at $Q^0 \approx 0.7$ kW to the values of $\gamma_L < 0.5$ (curve 1 in Figure 2.19). It means that the growth of Q^0 at low currents leads to an almost jump-like transition from mostly the arc method for maintaining the plasma in the region under consideration ($\gamma_L > 0.5$) to mostly the optical one ($\gamma_L < 0.5$). At high currents (curves 2 and 3 in Figure

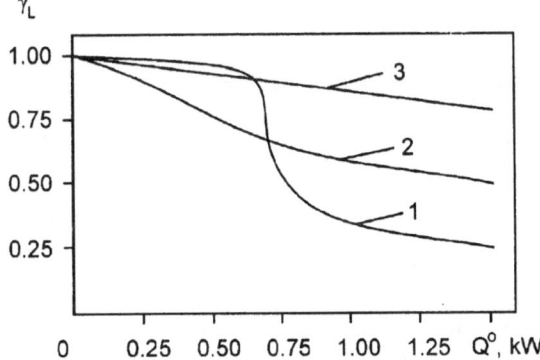

Figure 2.19 Relationships between parameter γ_L and the laser beam power at L = 2 cm for different arc currents: 1, 1 = 10 A; 2, 50 A; 3, 150 A (other parameters are the same as in Figure 2.10).

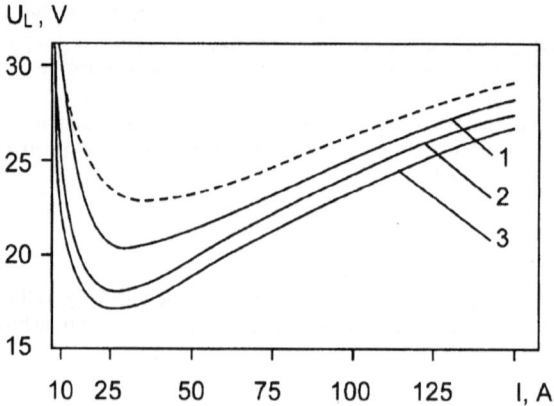

Figure 2.20 The voltage-current characteristics of the discharge at $L = 2$ cm for different Q^0
(parameters and designations are the same as in Figure 2.10).

2.19) the observed changes in the total energy balance of the plasma, depending on Q_0, are smoother, while the decrease in the discharge voltage under the effect of the laser radiation becomes less pronounced (see Figure 2.20). It should be noted that the calculated voltage-current characteristics plotted in Figure 2.20 for the column of the argon arc affected by the CO_2-laser radiation and the character of their variation with the growth of the laser beam power are in good agreement with the experimental data obtained by the authors of [54] for laser + GTA welding in an argon atmosphere (see Figures 1.33, 1.34). This points to the fact that the basic mechanism underlying a decrease in the argon arc voltage in combined welding using the CO_2-laser is a reduction of the field intensity in the column due to additional heating of the arc plasma by laser radiation, rather than a reduction of the anode potential drop due to evaporation of the anode material, as stated in [54].

   As a whole, the results of numeric modelling of the effect of focused CO_2-laser radiation on the plasma of the argon arc burning inside a long cylindrical channel suggest that the said effect at $I > 10$ A, $(R_C = 2.5$ mm) and $Q^0 > 0.7$ kW $(r_F = 0.2$ mm) induces a combined laser-arc discharge whose distributed and integral characteristics differ essentially from the corresponding characteristics of the conventional arc in the channel.

2.3.2 Propagation of the laser beam in the arc plasma

Consider the effect of the discharge plasma in argon, described above, on the focused beam of CO_2-laser radiation propagating in it. As follows from Figure 2.8, the refractive index $n_\omega = \sqrt{\varepsilon_\omega'}$ of the argon plasma and its absorptivity $\kappa_\omega = k\varepsilon_\omega'' / \sqrt{\varepsilon_\omega'}$ greatly depend on its temperature. Because of the sufficiently high heterogeneity of the distribution of T in the discharge under consideration (see, e.g. Figures 2.11, 2.16a and 2.18a) the optical properties of the plasma are also rather non-uniform in space.

   Considerable changes in the spatial distribution of the radiation intensity due to refraction and attenuation of the beam should be observed, if the laser beam propagates in such an optically heterogeneous absorbing medium. Indeed, this assumption is proved by calculations made. Thus, for example, Figure 2.21 illustrates the length variations of the laser radiation intensity $S_0(z)$ at the axis of the beam and its effective radius $r_b(z)$ determined from the relationship:

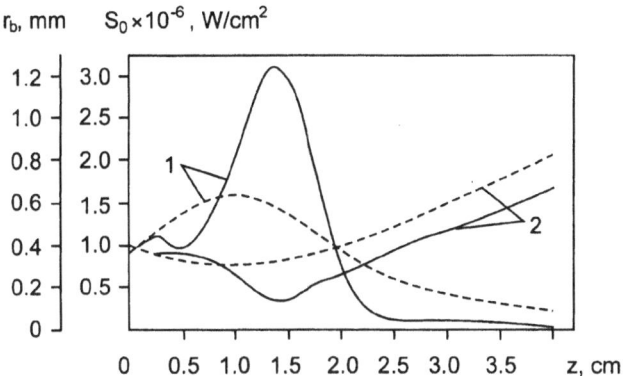

Figure 2.21 Variations of the laser radiation intensity on the beam axis and the effective radius of the laser beam along the channel (I = 100 A; R_C = 2.5 mm, R_1 = 1.0 mm; G = 0.1 g/s, G_1 = 0.01 g/s; Q^0 = 1.0 kW; r_F = 0.2 mm, F = 10 mm): 1, $S_0(z)$; 2, $r_b(z)$; dashed curves correspond to the beam without plasma.

$$S(r_b, z) = 0.01\, S_0(z). \tag{2.64}$$

The dashed lines in this Figure depict similar dependences for the initial beam at the absence of the plasma, plotted by formulae (2.14), (2.27) – (2.31), (2.33), (2.64). As seen from comparing the appropriate curves, at the very beginning of the laser-arc interaction region there occurs a slight broadening of the beam (see curves 2) which, combined with absorption of its power, leads to a marked decrease in the S_0 value (see curves 1). Defocusing of the laser beam observed here is caused by the fact that, when entering the plasma in the initial sections of region II, where the temperature already reaches a maximum on the discharge axis, but does not yet exceed 16500 K*, the beam propagates in the optically heterogeneous medium with a minimum value of ε_ω' (and, therefore, of n_ω) on its axis. It is known from geometrical optics that while propagating in a heterogeneous medium, the rays deflect towards an increase in the refractive index [110]. An increase in the axial temperature of the plasma above 16500 K occurring with growth of z, leads to the minimum of its permittivity in accordance with the T dependence of ε_ω' starting to shift from the axis towards the periphery of the discharge (curve 1 in Figure 2.22a). As a result, a region with $\partial \varepsilon_\omega' / \partial r < 0$ forms near the beam axis and plays the role of a convergent lens for the central beam rays, while the peripheral beam rays are still subjected to defocusing. Increasing in diameter with distance from the beginning of the laser-arc interaction region as a result of heating of the central plasma regions by laser radiation, the plasma lens (region with $\partial \varepsilon_\omega' / \partial r < 0$) gradually fills in the entire section of the beam (curves 2 and 3 in Figure 2.22a). This implies a transition to its intensive focusing over the entire section. Here, the effective radius r_b determined by relationship (2.64) decreases to such an extent (see curves 2 in Figure 2.21), that, despite considerable absorption of the beam in the plasma, the maximum power density S_{max} of the radiation at its axis grows up to a value which is almost twice as high as that in the focus of the initial beam (see curves 1 in Figure 2.21). Besides, the focus location f determined, for example, from the condition

$$S_0(f) = S_{max}, \tag{2.65}$$

---

* This temperature corresponds to the minimum value of ε_ω' for the argon plasma at atmospheric pressure (see curve 1 in Figure 2.8a).

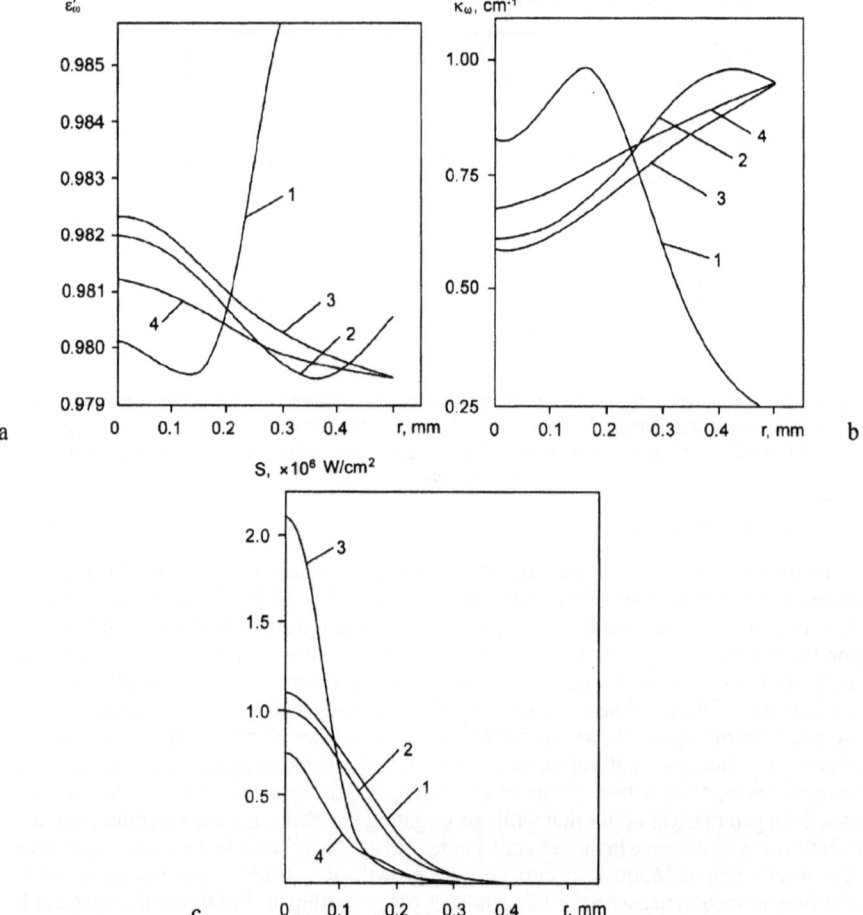

Figure 2.22 Radial profiles of real part of the complex permittivity (a), the absorptivity of the plasma (b) and the laser radiation intensity (c) for different cross-sections of the channel (parameters are the same as in Figure 2.21): 1, $z = 3$ mm; 2, 5 mm; 3, 10 mm; 4, 20 mm.

shifts along the beam to a distance of $\Delta_f = f - F \approx 4.0$ mm (see Figure 2.21). Since, upon reaching the maximum, the temperature of the plasma at the discharge axis starts slowly decreasing with the further increase in z, the focusing properties of the plasma lens gradually attenuate (curve 4 in Figure 2.22a).

To analyse changes in the radial distributions of the laser radiation intensity taking place during propagation of the beam in the laser-arc discharge plasma, one should take into account, in addition to the above effect of refraction of the beam, the essentially heterogeneous distribution of absorptivity κ_ω over its section (Figure 2.22b). Radial profiles of the laser radiation power in different sections of region II have the form shown in Figure 2.22c.

To estimate the relative role of the refraction and absorption processes in the beam propagation in the plasma of the studied discharge, we introduce coefficients of the laser

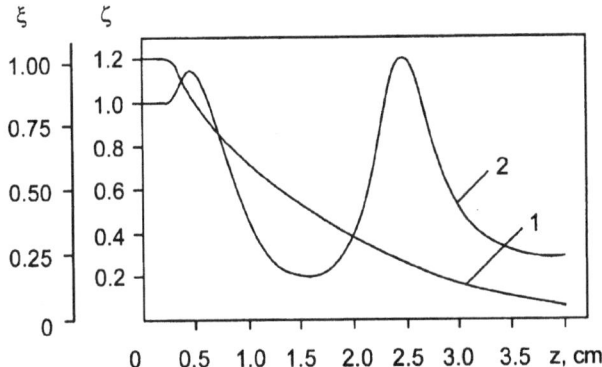

Figure 2.23 Longitudinal distributions of the beam attenuation and the refraction broadening coefficients (parameters are the same as in Figure 2.21): 1, $\xi(z)$; 2, $\zeta(z)$.

beam attenuation $\xi(z)$ and refraction broadening $\zeta(z)$ (see, e.g. [111]) which are determined by relationships:

$$\xi = \frac{Q}{Q^0};$$
(2.66)

$$\zeta = \frac{Q}{Q^0}\frac{S_0^0}{S_0},$$
(2.67)

where S_0^0 is the power density of the laser radiation at the beam axis in the absence of plasma. Distributions of the said coefficients along the discharge length are shown in Figure 2.23. Unlike the ξ value, which is a monotonically decreasing function of z (curve 1), the refraction broadening coefficient ζ behaves non-monotonically. Some broadening of the beam with respect to the initial one (region $\zeta > 1$ in curve 2) observed at the beginning of the laser-arc interaction region changes into intensive focusing (region $\zeta < 1$), the beam then broadens

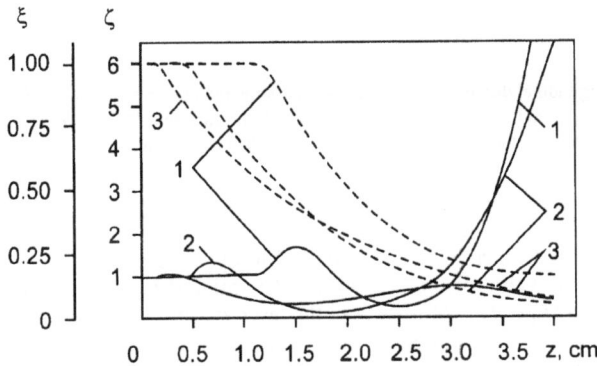

Figure 2.24 Longitudinal distributions of the beam attenuation and the refraction broadening coefficients for different arc currents: 1, $I = 10$ A; 2, 50 A; 3 150 A (other parameters are the same as in Figure 2.21); solid curves, $\zeta(z)$; dashed curves, $\xi(z)$.

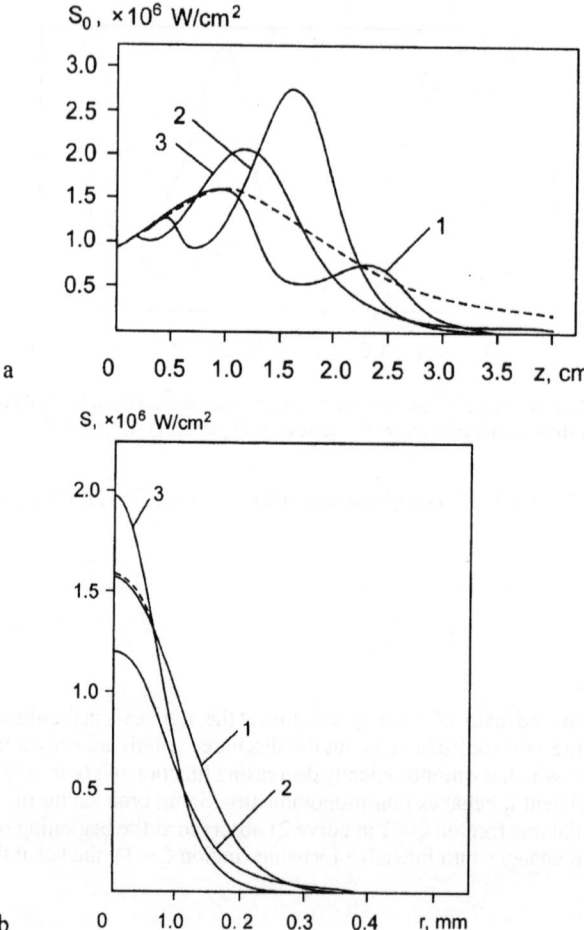

Figure 2.25 Longitudinal distributions of the laser radiation intensity on the beam axis (a) and radial distributions of the laser radiation intensity at $z = 10$ mm (b) for different I: 1, $I = 10$ A; 2, 50 A; 3, 150 A (other parameters are the same as in Figure 2.21); dashed curves, laser beam without plasma.

again ($\zeta > 1$) and, finally, stabilisation of its radius occurs with respect to the radius r_b^0 of the initial beam (see also curves 2 in Figure 2.21).

Due to the strong temperature dependence of the optical properties of the argon plasma (see Figure 2.8), the value of the arc current determining in many respects the spatial distribution of the temperature of plasma in the laser-arc discharge greatly affects also the characteristics of the laser beam propagating in it. Thus, for example, reduction of the region I length (see Figure 2.18) occurs with an increase in current and leads to such a situation. Absorption of the energy and refraction broadening of the beam which enters the plasma start first in z, but, become less intensive (Figure 2.24). Focusing of the beam in plasma observed after that also starts earlier. As a result, with the growth of current, a section with minimum width of the beam (i.e. the position of a second maximum of the S_0 value)

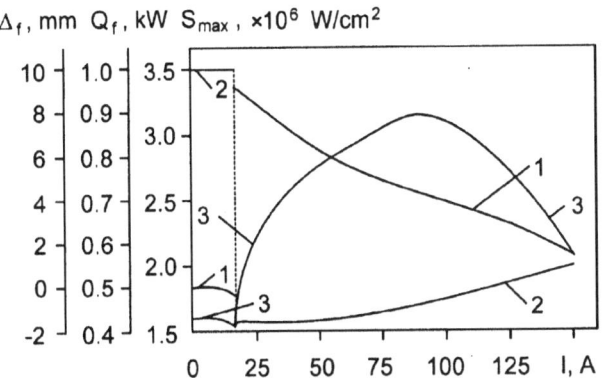

Figure 2.26 The focus shift Δ_f, the focus power Q_f and the maximum intensity of the laser radiation S_{max} as functions of I (other parameters are the same as in Figure 2.21): 1, $\Delta_f(I)$; 2, $Q_f(I)$; 3, $S_{max}(I)$.

approaches the plane of $z = F$. In the case of very low currents ($I = 10$ A) the second maximum of the radiation intensity, S_0, related to focusing of the beam by the plasma lens, is below the first one (curve 1 in Figure 2.25a) and the focus position determined from condition (2.65) practically coincides with the $z = F$ plane (left branch of curve 1 in Figure 2.26). At $I = 17$ A both maxima are equal in value, i.e. the beam in the plasma has, in fact, two focuses (gap in curve 1 in Figure 2.26), the second being much farther than the focal plane of the initial beam. The laser radiation power $Q_f \equiv Q(f)$ reaching the second focus is much lower than the corresponding value for the first focus (gap in curve 2 in Figure 2.26). With the growth of current ($I > 17$ A) a section, wherein the maximum value of the radiation intensity is achieved, approaches the $z = F$ plane (compare curves 2 and 3 in Figure 2.25a). Therefore, Δ_f monotonically decreases, while Q_f increases (right branches of curves 1 and 2 in Figure 2.26). As to the maximum intensity S_{max} of the laser radiation achieved at the beam axis, it first grows with increase in the arc current and then reduces (right branch of curve 3 in Figure 2.26), so the absolute maximum of the S_0 value for the selected parameters of the initial beam ($Q^0 = 1.0$ kW, $r_F = 0.2$ mm, $F = 10$ mm) takes place at $I \approx 90$ A. As far as distributions of the laser radiation intensity over the beam section in the discharge studied (Figure 2.25b) are concerned, they also greatly depend on the arc current and the selected section position (see also Figures 2.24, 2.25a).

Conditions for propagation of the laser beam in the combined discharge plasma depend not only on the arc current, but also on the pattern of the gas flow in the initial region of the channel. In particular, an increase in the length of region I with growth of G_1 (see Figure 2.17) leads to the fact that the maximum of the S_0 value shifts farther along z and decreases to some extent (Figure 2.27). An increase in Δ_f is also observed with an increase in the total gas flow G through the channel, whereas the maximum value of the radiation intensity at the beam axis practically does not change (compare dashed and solid curves in Figure 2.27). It raises the feasibility of gas dynamic control over focusing laser radiation in the laser-arc discharge plasma by changing the plasma gas flows G_1 and G_2.

Behaviour of the laser beam in the discharge under consideration is affected basically by its power. This is associated with the fact that with an increase in Q^0 there occurs a substantial redistribution of the plasma temperature in the discharge (see Figure 2.11) and, as a result, a change in spatial distributions of permittivity and absorptivity (Figure 2.28). In the case of low power of laser radiation ($Q^0 < 0.3$ kW) and $I < 100$ A ($R_C = 2.5$ mm) the

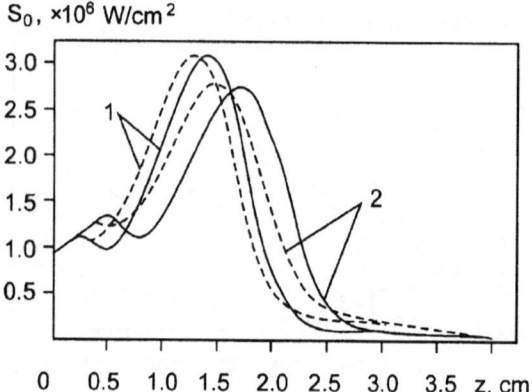

Figure 2.27 Distributions of the laser radiation intensity on the beam axis along the channel for different plasma gas flow rates: 1, $G_1 = 0.01$ g/s; 2, 0.025 g/s; solid curves, $G = 0.1$ g/s; dashed curves, 0.05 g/s (other parameters are the same as in Figure 2.21).

plasma temperature at the discharge axis does not go over 16500 K and the beam propagates in a medium the refractive index of which is a monotonically rising function of radius (see dashed curve in Figure 2.28a), i.e. undergoes a permanent broadening. At $Q^0 = 0.5$ kW, $I = 100$ A in the sections, where T exceeds 16500 K, a small region with $\partial \varepsilon_\omega' / \partial r < 0$ (see curve 1) starts forming near the system axis. However, due to the small values of $\partial \varepsilon_\omega' / \partial r$ and due to a growth of the laser radiation absorptivity of the plasma (see curve 1 and dashed curve in Figure 2.28b), the effect of the beam focusing in the plasma is not yet apparent. On the contrary, a significant broadening of the beam, as compared to the initial one (solid curve 1 in Figure 2.29), takes place in these sections due to defocusing of the peripheral rays, which, together with a decrease in the total power of the laser radiation while

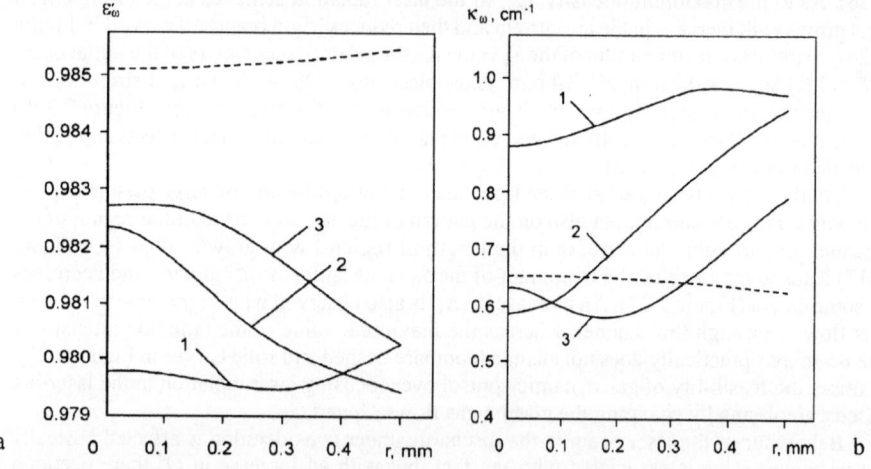

Figure 2.28 Radial profiles of real part of the permittivity (a) and the absorptivity (b) of the combined discharge plasma at $z = 10$ mm for different laser beam powers: 1, $Q^0 = 0.5$ kW; 2, 1.0 kW; 3, 1.5 kW; dashed curves, $Q^0 = 0$ (other parameters are the same as in Figure 2.21).

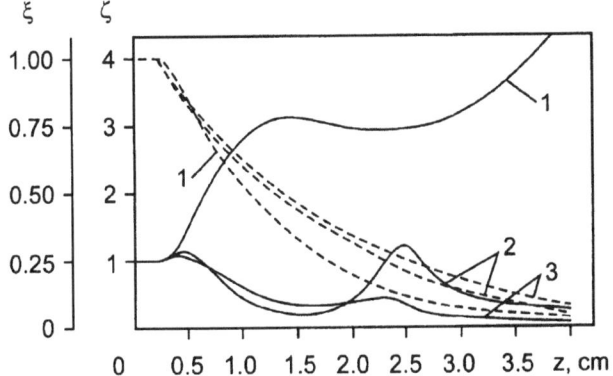

Figure 2.29 Longitudinal distributions of the beam attenuation and the refraction broadening coefficients for different Q^0: 1, $Q^0 = 0.5$ kW; 2, 1.0 kW; 3, 1.5 kW (other parameters are the same as in Figure 2.21); solid curves, ζ (z); dashed curves, ξ (z).

propagating along the discharge axis (dashed curve 1 in Figure 2.29), leads to a decrease in the radiation intensity at the axis (curves 1 in Figure 2.30).

With the growth of the initial beam power the focus position f, corresponding to the maximum S_0 value, shifts towards the beam. Besides, while $Q^0 < 0.73$ kW, Δ_f depends but slightly on Q^0 (left branch of curve 1 in Figure 2.31) and the power Q_f of the laser radiation reaching the focus and its maximum intensity S_{max} do not practically differ from the corresponding values for the initial beam in plane $z = F$ (left branches of curves 2 and 3 in Figure 2.31). Further increase in the laser beam power, causing a corresponding growth of the temperature within the near-axis zone of the discharge, results in an increase in ε_ω' and a decrease in κ_ω at the beam axis (compare curves 1 and 2 in Figures 2.28a, b). As a result, it is at $Q^0 = 1.0$ kW, after a certain broadening of the beam within the initial sections of the laser-arc interaction region, that the intensive focusing of the laser radiation in the plasma occurs (solid curves 2 in Figures 2.29, 2.30), while a decrease in the total power Q with growth

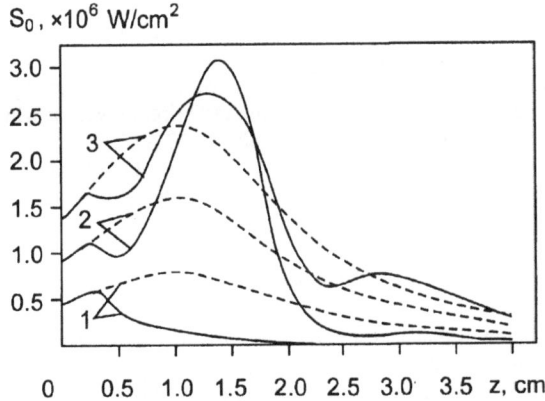

Figure 2.30 Longitudinal distributions of the laser radiation intensity on the beam axis for different Q^0: 1, $Q^0 = 0.5$ kW; 2, 1.0 kW; 3, 1.5 kW (other parameters are the same as in Figure 2.21); solid curves, laser beam in plasma; dashed curves. without plasma.

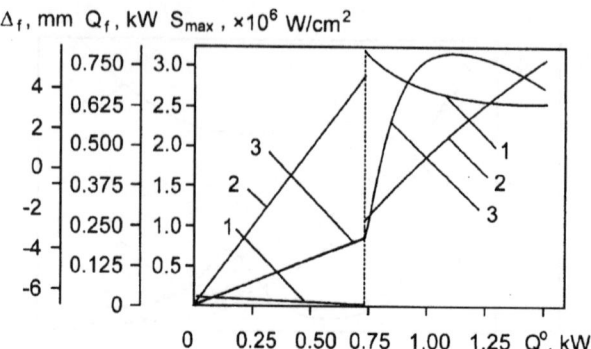

Figure 2.31 The focus shift Δ_f, the focus power Q_f and the maximum intensity of the laser radiation S_{max} as functions of Q^0 (other parameters are the same as in Figure 2.21): 1, $\Delta_f(Q^0)$; 2, $Q_f(Q^0)$; 3, $S_{max}(Q^0)$.

of z is slower (compare dashed curves 1 and 2 in Figure 2.29). It should be noted, that as soon as Q^0 becomes higher than 0.73 kW, the beam focus shifts in a jump-like manner outside the $z = F$ plane (a gap in curve 1 in Figure 2.31), whereas the value of Q_f drastically drops (a gap in curve 2 in Figure 2.31). A further rise of Q^0 leads to some decrease in Δ_f and to almost linear growth of Q_f (right branches of curves 1 and 2 in Figure 2.31). The maximum intensity S_{max} of the radiation in the beam focus first grows with an increase in Q^0 and, then (at $Q^0 > 1.1$ kW), decreases (curve 3 in Figure 2.31). It means that with growth of the power of the laser beam above the said value, its additional focusing in the plasma becomes somewhat lower (compare solid curves 2 and 3 in Figures 2.29, 2.30).

Therefore, the results of numerical study of the propagation of the CO_2-laser radiation beam in the plasma of an argon arc burning inside the channel, show that in such a system, along with the laser radiation absorption effect, some other optical effects can take place as well, which must be taken into account when analysing the influence of the arc plasma on the beam in real combined processes. In particular, redistribution of optical properties of the arc plasma under the effect of the laser radiation leads to the formation of a plasma lens induced by certain values of the arc current and the beam power, which provides additional focusing of the laser radiation. Note, that since the said effect is of a threshold character for the laser beam power, it could not be detected, for example, by experimental studies on low-power laser beams passing through the arc plasma, as conducted by the authors of [60]. The threshold values of the arc current and the radiation power, at which self-focusing of the laser beam in plasma occurs, are, respectively: $I = 17$ A ($R_C = 2.5$ mm) and $Q^0 = 0.73$ kW ($r_F = 0.2$ mm), for the conditions considered in the present work, i.e. lie within the range of the parameters characteristic particularly of the combined discharge.

2.3.3 Influence of plasma gas on the characteristics of the laser-arc discharge

In practical realisation of the laser-arc discharge it may be necessary to use not only Ar as a plasma gas, but also some other inert gases, for example, He or its mixtures with Ar. Therefore, it seems reasonable to study dependence of the basic characteristics of the combined discharge plasma and the CO_2-laser radiation beam interacting with it on the composition of the plasma-forming medium.

Consider a discharge inside the channel blown through with a flow of Ar – He mixture with various volume ratios of stock components. The geometrical dimensions of the channel

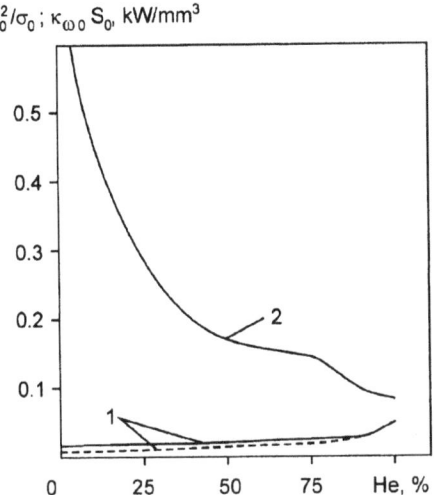

j_0^2/σ_0 ; $\kappa_{\omega 0}\,S_0$, kW/mm^3

Figure 2.32 Effect of He proportion on the axial values of the heat sources in Ar – He discharge plasma at z = 10 mm (I = 100 A; R_C = 2.5 mm, R_1 = 1.0 mm; G^V = 5.0 l/min, G_1^V = 0.5 l/min; r_F = 0.2 mm, F = 10 mm): 1, j_0^2/σ_0; 2, $\kappa_{\omega 0}S_0$; solid curves, Q^0 = 1.0 kW; dashed curve, Q^0 = 0.

and the cathode nozzle (see Figure 2.1) are the same as in Item 2.3 a, the pressure at the channel exit is atmospheric, volume flow rates of the plasma-forming mixture are G^V = 5.0 l/min, G_1^V = 0.5 l/min*. The arc current is 100 A, and the laser radiation power is Q_0 = 1.0 kW. The distance between the initial beam focusing plane and the inlet channel section is F = 10 mm, and the beam radius in this plane is r_F = 0.2 mm.

As noted above, the basic mechanism of the laser effect on the arc is additional heating of the arc plasma by laser radiation. The energy put into the plasma by the laser beam (source term $\kappa_\omega S$ in equation (2.1)) is a function of both the laser radiation intensity and the coefficient of its absorption by the plasma. Because κ_ω greatly depends on the plasma composition (2.57), a change in the ratio of the stock components in the plasma-forming mixture under consideration should lead to corresponding changes in the energy balance of the combined discharge, other conditions being equal.

Figure 2.32 shows the axial values of the sources of heating the plasma by the current flow and the absorbed laser radiation, depending on the helium percentage of the plasma-forming mixture, characteristic of the laser-arc interaction region. An increase in the He volume fraction causing a decrease in the laser radiation absorptivity leads to a lower role played by the laser beam in the energy balance of the discharge (see curve 2). The role of Joule heating of the plasma, on the contrary, grows (see solid curve 1), this being in agreement with the well-known fact of a rise in the electric field intensity and the current density in the Ar – He arc column with growth of the helium content of the mixture [112]. Eventually, affected by laser radiation, the change in the Joule source j_0^2/σ_0 (on the discharge axis) is much less in helium than in argon (see dashed and solid curves 1). Hence, it can be suggested, that the effect of the CO_2-laser beam radiation on the Ar – He arc plasma characteristics will diminish with increase in the He content of the plasma-forming mixture.

---

* Under constant pressure and temperature, the density of the Ar – He mixture greatly depends (see Figure 2.2) on the volume fraction of He, i.e. on the parameter δ (2.41). Hence the most convenient criterion for the plasma gas flow rate is the volume flow rate, rather than the mass flow rate.

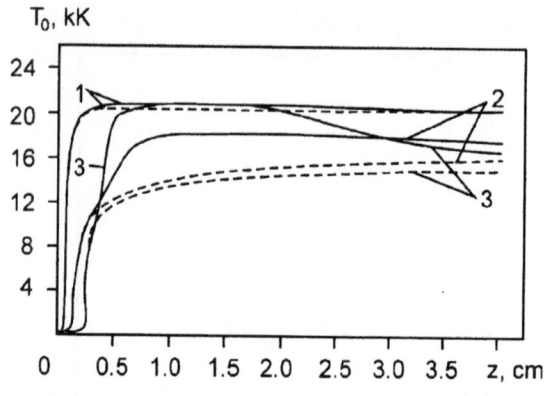

a

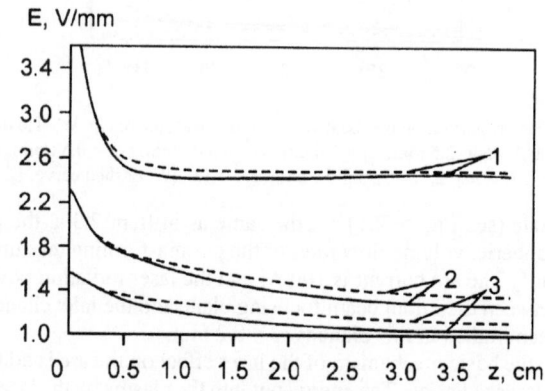

b

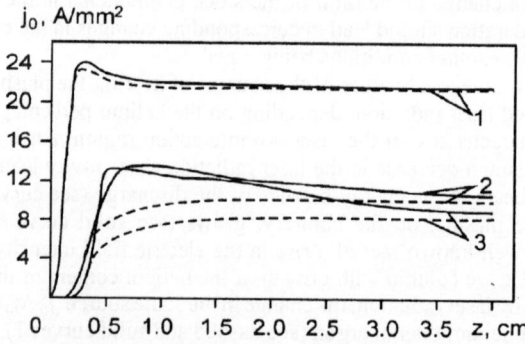

c

Figure 2.33 Distributions of the plasma temperature on the discharge axis (a), the electric field (b), the axial current density (c), the axial velocity of plasma (d) and the parameter γ (e) along the channel for different compositions of plasma gas: 1, 100 % He; 2, 50 % Ar + 50 % He; 3, 100 % Ar; solid curves, $Q^0 = 1.0$ kW; dashed curves, $Q^0 = 0$ (other parameters are the same as in Figure 2.32).

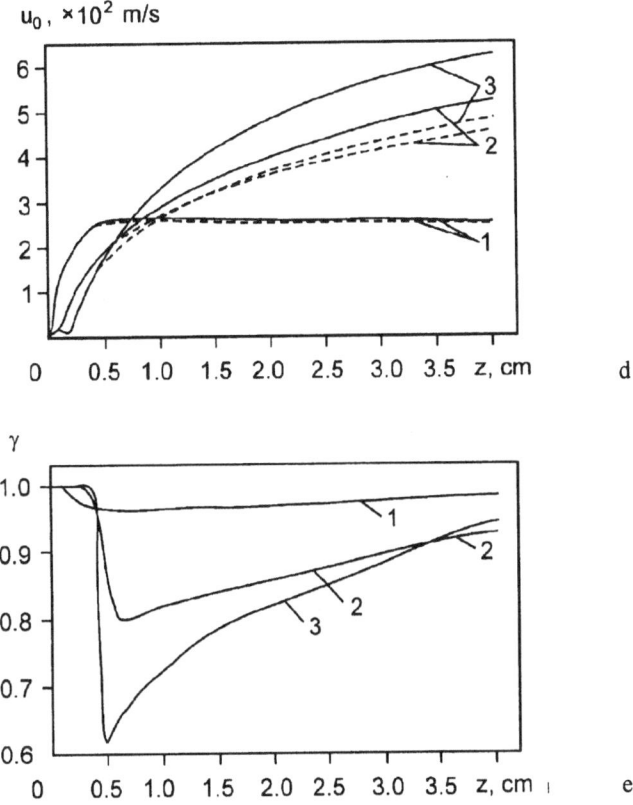

Figure 2.33 (*continued*).

Actually, the relative elevation of the axial temperature of the arc plasma under the effect of laser radiation against the significant growth of T_0 with lower Ar content of the plasma-forming mixture, becomes less and less essential with its enrichment with helium (Figure 2.33a). As a result, at $I = 100$ A, $Q^0 = 1.0$ kW, the maximum achievable value of the laser-arc discharge plasma temperature in Ar is even a bit higher than in He (compare solid curves 1, 3). Along with changes in the absolute values of the plasma temperature at the discharge axis with a change in the plasma composition, there occurs a marked redistribution of T_0 along the channel length. An increase in the percentage of helium having the higher thermal conductivity (see Figure 2.5) results in the fact that the annular flow of the arc plasma formed by the cathode nozzle collapses closer and closer to the inlet section of the channel, i.e. region I becomes shorter (see Figure 2.33a). Because of a decrease in the absorption of the beam in the plasma the laser-arc interaction region (region II), on the contrary, becomes longer, while lowering of T_0 in the conversion region becomes slower.

The relative variation in the electric field intensity occurring in the discharge under the effect of the laser radiation depends but slightly on the plasma gas composition and is negligible, compared with increase in E with growth of the He content of the arc plasma (Figure 2.33b). Current density at the channel axis, like in the conventional arc discharge, rises with growth of the helium concentration of the mixture (Figure 2.33c). In this case the relative increase in j_0 due to heating the plasma by the laser beam becomes less and is less

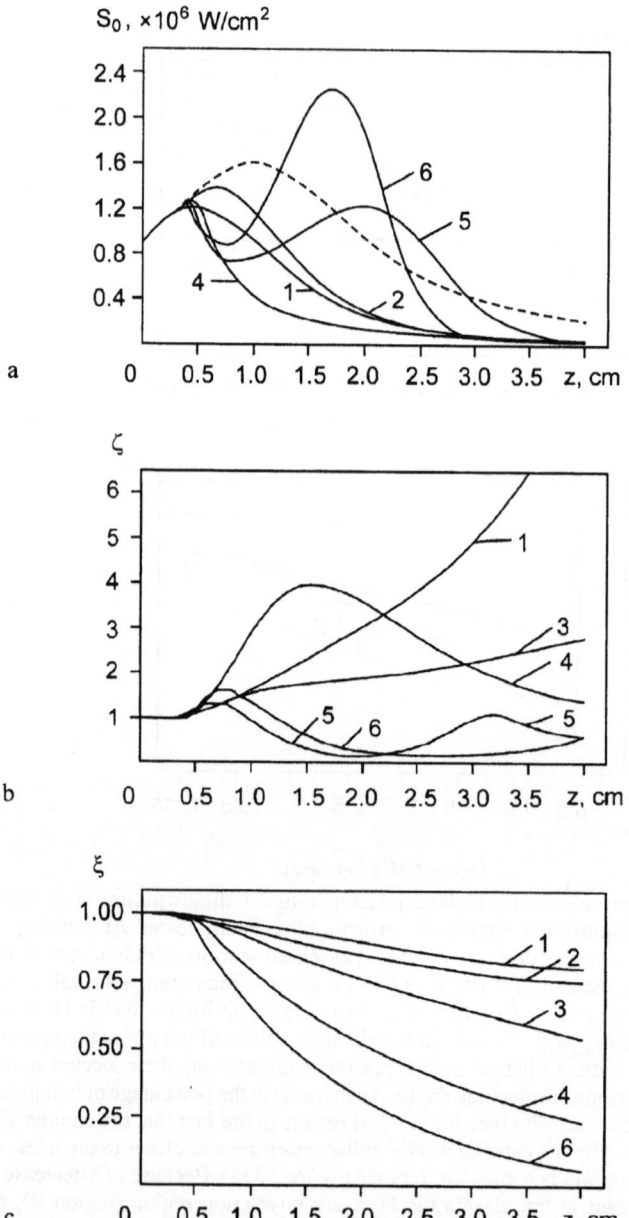

Figure 2.34 Longitudinal distributions of the laser radiation intensity on the beam axis (a), the refraction broadening (b) and the beam attenuation coefficient (c) for different compositions of plasma gas: 1, 100 % He; 2, 5 % Ar + 95 % He; 3, 25 % Ar + 75 % He, 4, 50 % Ar + 50 % He; 5, 90 % Ar + 10 % He; 6, 100 % Ar (other parameters are the same as in Figure 2.32); solid curves, laser beam in plasma; dashed curve, without plasma.

essential. As to a change in the j_0 distribution along the channel length depending on the composition of the plasma gas, it is similar to the one described above for the distribution of the axial temperature of the laser-arc discharge plasma.

A less significant role of the laser beam is played in the energy balance of the discharge considered with an increase in the He content of the plasma-forming mixture and leads to a smaller effect of the laser radiation on the hydrodynamic characteristics of the plasma flow as well, as illustrated in Figure 2.33d. Dependences of $u_0(z)$ plotted in this Figure allow the conclusion to be drawn that the length, at which the asymptotical distribution of the plasma velocity is established, diminishes with the growth of the helium content of the mixture. In this case the value of the axial plasma velocity in the asymptotical region of the channel drops due to increase in the mixture viscosity (Figure 2.4).

Variations in the axial distributions of the laser-arc discharge plasma characteristics inside the channel caused by variations in the plasma gas composition can be compared with the behaviour of the parameter γ (Figure 2.33e) characterising, according to (2.60), the relative contributions of the arc (current) and laser heating of the plasma to the total energy balance of the discharge. As follows from comparing curves 1–3, an increase in the helium concentration of the gas mixture causes a decrease in the role played by the laser beam in the energy balance of the discharge considered, i.e. an attenuation of the laser effect on the arc plasma. Of special note is the fact, that in pure He the region of the marked laser-arc interaction determined, e.g., from relationship $\gamma < 0.9$, is absent, since the laser beam contributes not more than 3 % of the total power supplied to the plasma (see curve 1). This implies that it is impossible to realise the combined discharge in helium at $I = 100$ A ($R_C = 2.5$ mm) and $Q^0 = 1.0$ kW ($r_F = 0.2$ mm). As shown by earlier calculations made within the framework of one-dimensional model [65], the $Q^0 > 3.0$ kW power of the CO_2-laser radiation focused to the $S_{max}^0 > 10^6$ W/cm^2 intensity values is required to realise the laser-arc discharge in He at the channel radius of $R_C = 2.5$ mm and the arc current of $I = 125$ A.

A change in the plasma gas composition influences not only the degree of laser radiation effect on the arc, but also leads to fundamental change in the conditions for propagation of the beam in the arc plasma and, hence, in its characteristics, this being illustrated in Figure 2.34. Thus, for example, only 10 % He added to argon causes a noticeable decrease in the effect of the beam self-focusing (compare curves 5, 6 in Figure 2.34a). With the increase in the He volume fraction of up to 50 % this effect totally disappears and further on the laser beam undergoes a gradual broadening in the plasma, aggravating with the helium enrichment of the plasma-forming mixture (see curves 1, 3, 4 in Figure 2.34b). The cause of defocusing of the laser radiation in the helium-rich mixtures ($\delta > 0.5$) is that, starting from the said He concentration, the values of maximum temperature over the channel section reached at the axis of the discharge under consideration (see Figure 2.33a) do not go outside the bounds of the drooping region for the corresponding dependence of $\varepsilon_\omega'(T)$ (see Figure 2.8a), i.e. the laser beam propagates in the medium with the minimum of ε_ω' and, hence, n_ω at its axis.

As to attenuation of the beam in the arc plasma, an increase in the He content of the plasma-forming mixture accompanied by a decrease in κ_ω should lead to a slower attenuation of the laser beam. Indeed, increase in the He volume fraction of up to 90 %, the rate of variation in the $\xi(z)$ parameter characterising the decrease in the power of laser radiation with its propagation along the arc axis greatly decreases (compare curves 3, 4, 6 in Figure 2.34c). However, further growth of the He percentage, on the contrary, leads to a certain aggravation of the beam attenuation. As a result, in pure He the attenuation is higher than, for example, in the 5 % Ar + 95 % He mixture (see curves 1, 2, respectively, in Figure 2.34c). Variation of the absorbing properties of the arc plasma caused by varying its composition accordingly affects the spatial distribution of the laser radiation intensity along the discharge burning in the mixtures rich in helium (see curves 1, 2, 4 in Figure 2.34a). However, it does not affect explicitly the character of the $S_0(z)$ dependence when using mixtures with a low content of He (see curves 5, 6 in Figure 2.34a), where, as noted above, redistribution of the radiation intensity occurs mainly due to a change in the focusing properties of the plasma. Therefore, the relative roles of the absorption and refraction processes in the formation of the spatial distribution of the radiation intensity and other characteristics of the laser beam interacting with the arc plasma greatly depend on its composition.

In general, results of a numerical study of the discharge induced by the effect of the focused CO_2-laser radiation on the Ar – He arc column plasma inside the channel suggest an efficiency of control over the laser-arc interaction processes in such a system by varying a ratio of the stock components in the gas mixture. In other words, varying the plasma gas composition can provide the optimum (from the standpoint of a particular technological process) spatial distributions of characteristics of the laser-arc discharge plasma and of the laser beam interacting with it. One should bear in mind here, that the regularities of behaviour of the combined discharge with variation of the plasma-forming medium will depend on particular conditions for its realisation, i.e. laser radiation power, arc current, channel dimensions, etc.

2.3.4 Peculiarities of the laser beam and the arc plasma interaction outside the channel

Consider a combined discharge forming in the interaction of the focused CO_2-laser radiation with the open arc column plasma in the axial argon flow. The initial region of the discharge is formed by the channel (see Figure 2.1) having a length of $L_C = 1.5$ cm, the radius of $R_C = 4.0$ mm and is blown through with argon with a mass flow rate of $G = 0.3$ g/s. The radius of the cathode nozzle is $R_1 = 1.5$ mm, and the rate of the gas flow through its internal cavity is $G_1 = 0.075$ g/s, the temperature of the cooled walls of the channel and the nozzle is $T_C = T_1 = 300$ K. The Gaussian radiation beam focused by an optical system is introduced into the discharge along the channel axis. The distance from the focal plane of the initial laser beam to the channel exit section is 5 mm ($F = 20$ mm), and the radius of the beam in the focusing plane is $r_F = 0.2$ mm. The laser-arc discharge in the region outside the channel is stabilised by the atmospheric pressure argon co-flow having the temperature $T_{ext} = 300$ K and the velocity $u_{ext} = 2$ m/s (see Figure 2.1). The length of the calculation domain is $L = 6$ cm, and its radius outside the channel is $R = 15$ mm. As shown by calculations, within the considered ranges of the discharge burning conditions ($50 A \leq I \leq 250 A, 0 \leq Q^0 \leq 1.5$ kW) the preset value of L is sufficient for manifestation of the peculiar behaviour of the plasma in all three regions characteristic of the combined discharge, and the selected value of R satisfies the conditions for smooth contingency with the environment (2.19).

The results of numerical modelling of the laser-arc discharge are given in Figures 2.35–2.42. As well as for the case considered earlier of the arc limited by the channel wall, the effect of the laser beam on the plasma of the open arc discharge region leads to local (near the column axis) elevation of the plasma temperature (see Figure 2.35a) caused by additional heating of the arc plasma by the laser radiation. A drastic growth of the axial temperature at the very beginning of the laser-arc interaction region (see Figure 2.35b) is replaced by a gradual decrease in T_0 in the conversion region down to the value characteristic of the non-disturbed arc in the gas flow and is accompanied by appropriate restructuring of the radial plasma temperature profiles (see Figure 2.36). Therefore, when the CO_2-laser radiation beam propagates in the plasma of the open argon arc, a high-temperature region tightly coupled with the beam axis is formed in it. The temperature of this region increases with the growth of the laser radiation power, as is the case for the wall-stabilised discharge (compare Figures 2.35, 2.11).

An increase in the electric conductivity of the plasma within the high-temperature region, that forms in the open part of the arc discharge, leads to a decrease in the electric

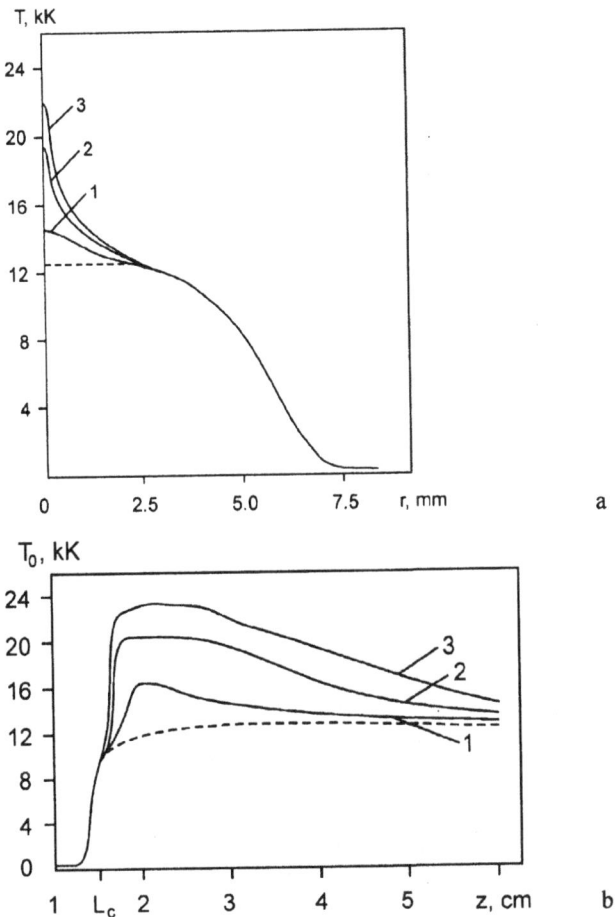

Figure 2.35 Radial profiles of the plasma temperature at z = 30 mm (a) and longitudinal distributions of the temperature on the discharge axis (b) outside the channel for different laser beam powers (I = 150 A; R_C = 4.0 mm, R_1 = 1.5 mm; L_C = 1.5 cm; G = 0.3 g/s, G_1 = 0.075 g/s; u_{ext} = 2.0 m/s; r_F = 0.2 mm, F = 20 mm): 1, Q^0 = 0.5 kW; 2, 1.0 kW; 3, 1.5 kW; dashed curves, Q^0 = 0.

field intensity in the corresponding sections of the arc column under the effect of the laser radiation (see Figure 2.37). However, due to a low ratio of transverse sizes of the zone of thermal disturbance caused in the plasma by the laser beam to the radius of the open arc column (as compared to a similar parameter for the arc inside the channel (compare Figures 2.35a and 2.11a)) the relative decrease in E is less pronounced (compare Figures 2.37 and 2.12). Because of this, the axial current density determined by the product σE grows in the open arc under the effect of the laser radiation to values higher than in the wall-restricted arc (compare Figures 2.38 and 2.13). Note that the radius of the current-conducting zone of the open discharge region does not practically change under the laser radiation effect on the arc plasma. Instead, it causes a redistribution of j over the section of the discharge (see Figure 2.38a). It should be noted too, that despite significant changes in the spatial distribution of

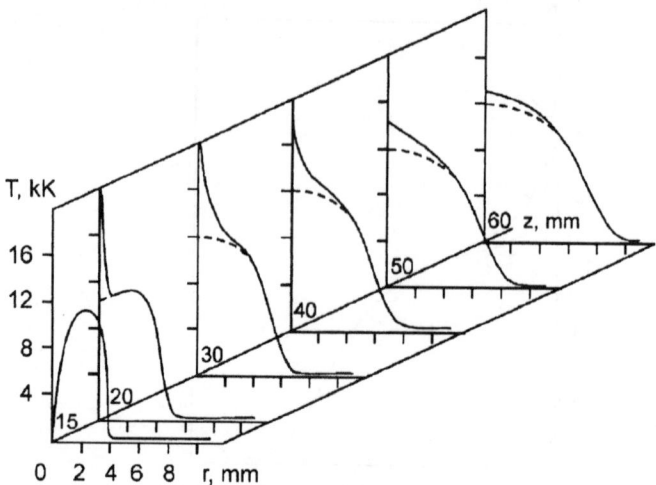

Figure 2.36 Radial distributions of the plasma temperature for different cross-sections of the open region of the discharge ($I = 100$ A; other parameters are the same as in Figure 2.35); solid curves, $Q^0 = 1.0$ kW; dashed curves, $Q^0 = 0$.

the current density in the open arc and, accordingly, the Lorentz forces acting on the plasma, no essential redistribution of its hydrodynamic characteristics occurs under the effect of the laser radiation within the considered range of the change in Q^0.

As in the case of the discharge inside the channel, the plasma temperature and the current density in the open laser-arc discharge depend not only on the laser beam power (see Figures 2.35, 2.38), but also on the arc current. The character of variations in the spatial distributions of the said characteristics with the current is similar to that described in item 2.3 a for the wall-stabilised discharge (compare Figures 2.39 and 2.18). Thus, a decrease in I leads to the fact that the annular plasma flow formed by the cathode nozzle collapses at a larger distance from the channel inlet section, i.e. region I becomes longer (see, e.g., Figure 2.39a). At $I <$

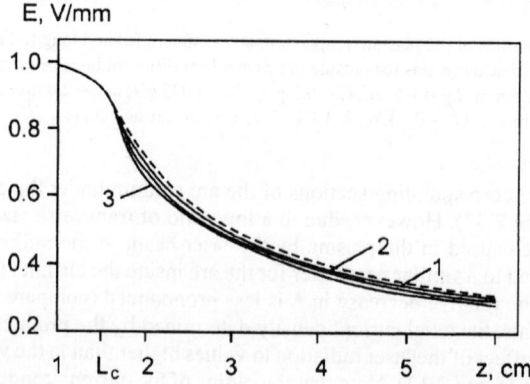

Figure 2.37 Distributions of the electric field along the axis of the open region of the discharge for different Q^0 (parameters and designations are the same as in Figure 2.35).

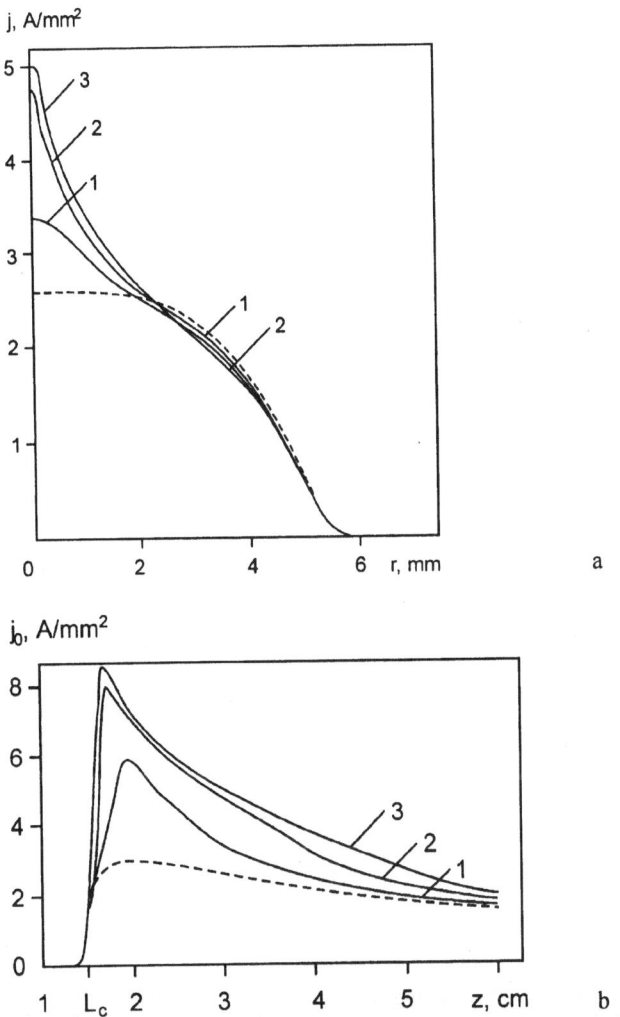

Figure 2.38 Radial profiles of the current density at $z = 30$ mm (a) and longitudinal distributions of the axial current density (b) in the discharge region outside the channel for different Q^0 (parameters and designations are the same as in Figure 2.35).

100 A the collapse takes place outside the channel and, hence, the entire region of laser-arc interaction is in the open part of the discharge. The relative increase in the axial values of temperature and current density in the discharge under the effect of the laser radiation becomes more significant with a decrease in the arc current, while the rate of lowering of T_0 and j_0 in the conversion region becomes higher (see Figure 2.39).

It should be noted, when considering the conditions for propagation of the laser beam in the plasma of an open region of the discharge, that in this case, as in interaction with the arc inside the channel, the beam is subjected to a strong effect of the arc plasma. In particular, the processes of refraction of the laser radiation in plasma, the role of which turns out to be

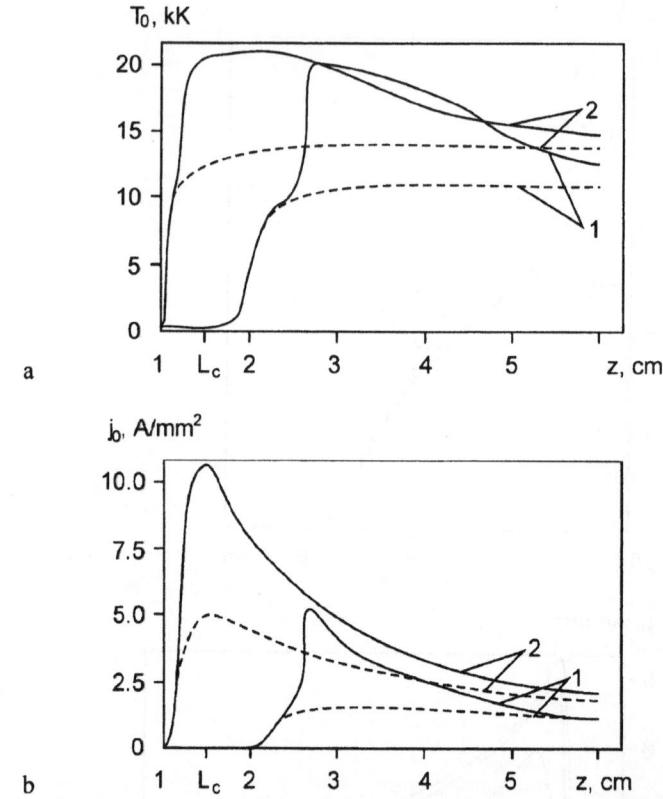

Figure 2.39 Longitudinal distributions of the plasma temperature on the discharge axis (a) and the axial current density (b) in the open region of the discharge for different I: 1, $I = 50$ A; 2, 250 A; solid curves, $Q^0 = 1.0$ kW; dashed curves, $Q^0 = 0$ (other parameters are the same as in Figure 2.35).

decisive in the initial section of the laser-arc interaction region, first cause some broadening of the beam and, then, its focusing. This then exerts a corresponding effect on the distribution of the radiation intensity in these sections (see Figure 2.40). At a distance from the point of introduction of the laser beam into the plasma, the effect of the absorption processes becomes stronger, leading to lower radiation intensity over the entire beam section.

The effect of self-focusing of the laser beam in the plasma of the open discharge region is similar to the case of the combined discharge burning inside the channel, and has a threshold character as to the laser radiation power. At the threshold value of Q^0, a sudden shift of the focal plane of the beam in the direction of its propagation ($\Delta_f > 0$) results. The only difference here is that under the conditions considered for realisation of the laser-arc discharge, the dependence of the maximum radiation intensity in a new focus of the laser beam on its power grows monotonically, while the shift of the focus becomes more significant (compare Figures 2.41, 2.30). The value of this shift Δ_f, as for the discharge inside the channel, does not practically depend on the laser radiation power and decreases with growth of the arc current (see Figures 2.41, 2.42). As to the maximum achievable radiation intensity in the laser beam interacting with the open arc discharge, its value, unlike

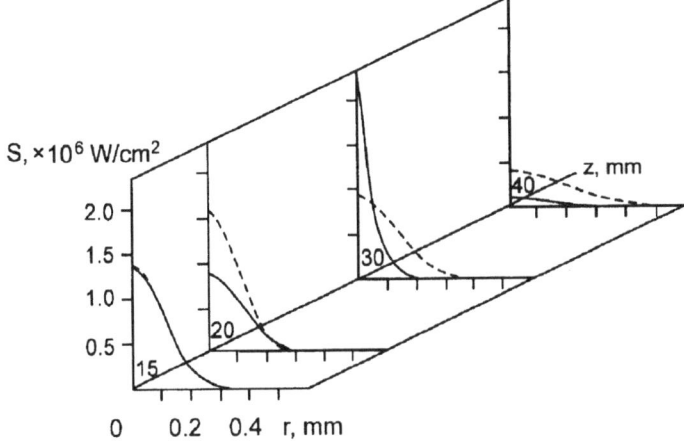

Figure 2.40 Radial distributions of the laser radiation intensity for different cross-sections of the open region of the discharge ($Q^0 = 1.0$ kW; other parameters are the same as in Figure 2.35); solid curves, laser beam in plasma; dashed curve, without plasma.

the discharge inside the channel, increases with the growth of the current over the entire range of its variation (compare Figures 2.42, 2.25a).

It should be noted, while completing consideration of the laser-arc discharge stabilised by a gas flow, as compared to the combined discharge stabilised by the channel wall, that regularities of interaction between the laser beam and the arc plasma are the same from the qualitative point of view. The differences in their quantitative character are related, first of all, to a change in the characteristics of the arc plasma proper, depending on the method of the arc column stabilisation, and point only to a different degree of the mutual effect of the arc and the plasma in the laser-arc discharge formed on the basis of either open or channel

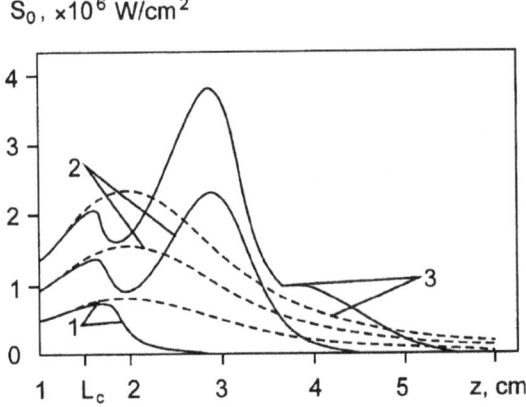

Figure 2.41 Longitudinal distributions of the laser radiation intensity on the beam axis in the open region of the discharge for different Q^0: 1, $Q^0 = 0.5$ kW; 2, 1.0 kW; 3, 1.5 kW (other parameters are the same as in Figure 2.35); solid curves, laser beam in plasma; dashed curves, without plasma.

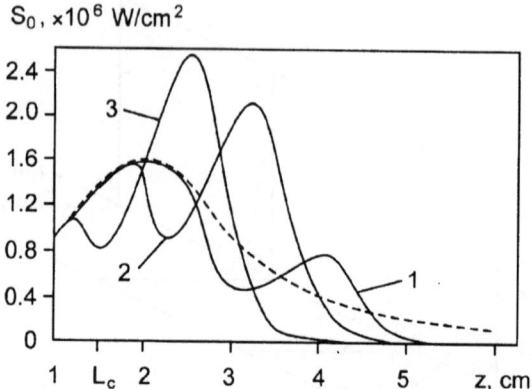

Figure 2.42 Longitudinal distributions of the laser radiation intensity on the beam axis for different I ($Q^0 = 1.0$ kW; other parameters are the same as in Figure 2.35): 1, I = 50 A; 2, 100 A; 3, 150 A; dashed curve, laser beam without plasma.

wall constricted arc. Hence, it can be concluded, that selection of such conditions for realisation of the combined discharge as a method for stabilisation and the degree of the arc column constriction is an efficient means for achieving the required technological properties of the laser-arc discharge plasma and the laser beam interacting with it.

CHAPTER 3

Integrated Plasma Torches for Laser-Arc Processes

Analysis of the results of numerical investigation into interaction of laser radiation with electric arc plasma given in the previous chapter shows that a special type of gas discharge, i.e., the combined laser-arc discharge, can exist in the "electric arc—laser beam" system. Commensurability of the energy introduced into the arc plasma by laser radiation with the energy released because of the electric current flowing in it is a necessary condition for formation of this discharge. If this condition is satisfied, a substantial change in the energy balance of the arc discharge occurs, which results in the characteristics of the combined discharge plasma and the initial arc plasma being fundamentally different*. Therefore, the laser-arc discharge enables low-temperature plasma with special properties to be generated and can be used as a basis for the development of a new class of plasma devices, i.e. integrated laser-arc plasma torches for hybrid laser-plasma processes of joining and treatment of materials.

The purpose of this chapter is to define the conditions required for practical realisation of the combined discharge and describe the basic arrangements and methods for calculation of laser-arc plasma torches of various applications . The results of mathematical modelling, experimental examination and technological application of the integrated plasma torch for laser-plasma surfacing are given.

3.1 Methods and Devices for Practical Realisation of the Combined Discharge

When realising the combined discharge in practice by way of the laser effect on the arc, the question arises as to the correct selection of the necessary parameters of the laser beam, the arc discharge mode and the method of combining the beam and the arc. Since the energy supplied by the laser radiation to the arc plasma in the laser-arc interaction region, depends on the power of the laser beam and the absorbing properties of the arc column plasma, determined in turn by the arc burning conditions, the above question can be solved in two ways. On the one hand, the power and other parameters of the laser beam can be selected on the basis of the preset arc discharge mode. On the other hand, the arc mode can be selected in accordance with characteristics of the available laser beam. If a combined discharge is created on the basis of the arc having a certain preset power, the power of the initial beam required for its formation should be determined from the coefficient of absorption of the laser radiation with a selected wave length in the given arc plasma. For example, when using the Nd: YAG-laser, the required power of the radiation should be tens of times higher than that when using the CO_2-laser, since radiation with $\lambda = 1.06$ micron wave length is absorbed by plasma approximately by two orders of magnitude lower (other conditions being equal), than radiation with $\lambda = 10.6$ microns wave length (see Figure 2.7). When selecting the initial

---

* Provided that the laser beam is capable of maintaining plasma when the current is switched off, i.e. if a continuous optical discharge can exist in the system under consideration, the combined laser-arc discharge can be interpreted also as the result of the electric current passing through the laser plasma. In this case the necessary condition for the formation of the combined discharge is to have commensurate energies supplied by the electric current to the optical discharge plasma and put into it by the laser radiation. Accordingly, it suggests significant differences between characteristics of the laser-arc discharge plasma and those of the initial laser plasma.

laser beam power one should take into account the fact that the coefficient of absorption of laser radiation in the arc column plasma depends not only on the radiation wave length, but also on the arc plasma temperature, pressure and the kind of gas used (see Figure 2.8b). Because of the latter circumstance, to realise the laser-arc discharge in slightly absorbent gases, e.g. in He, the power of the laser beam should be higher than in Ar. If a reverse problem is to be solved, namely: the laser radiation parameters are assigned, and the arc burning conditions necessary for combined discharge formation are to be determined, one should proceed from the fact that the temperature of the arc plasma, being one of the factors determining its absorbing properties, depends on the arc current, method of stabilisation and degree of constriction of the arc, as well as on the kind and pressure of the plasma-forming medium. Therefore, the above conditions for arc burning should be selected such that the plasma temperature in the path of the laser beam provides the absorption level necessary for realisation of the combined discharge. For example, in the case of using a CO_2-laser radiation beam with a power of approximately 1 kW and an atmospheric-pressure argon arc to form the laser-arc discharge, the temperature of the initial arc plasma should be not lower than 11000 K.

A highly efficient method for elevation of the arc column plasma temperature is constriction of the arc by restricting the transverse size of its column by the channel wall. This method makes it possible not only to achieve quite easily the formation of arc plasma with the temperature required for the combined discharge to form, but also to control this temperature and, consequently, the absorbing properties of the plasma by changing the arc current, the flow rate, the pressure and the composition of the plasma-forming gas. This will make it possible, by varying the said arc burning conditions, to effectively influence characteristics of the combined discharge plasma and, what is most important, characteristics of the laser beam interacting with it. Therefore, the channel stabilised electric arc used to form the laser-arc discharge, instead of the free-burning arc, will widen the range of the combined discharge burning conditions (by the arc current and the laser radiation power) and the possibilities of controlling its characteristics.

The most rational scheme for realisation of the combined discharge on the basis of the constricted (plasma) arc is that considered in the previous chapter involving the introduction of the laser beam into the arc column plasma along the axis of the stabilising channel. On the one hand, such a scheme suggests that the laser beam should travel in the arc plasma over a relatively large distance for ensuring the energy contribution of the laser radiation into the plasma, sufficient to form the laser-arc discharge. On the other hand, when using the combined heat source realised following such a scheme in welding, cutting or surface treatment of materials, in contrast to the method of laser-plasma welding [30, 113], the required coaxiality of the effect of the laser beam and the arc on the workpiece surface is ensured. It should be noted here that laser + PA welding, cutting and surface heat treatment (i.e. the processes using the combined discharge formed on the base of the constricted arc), while possessing all the advantages of the known laser + GTA processes, have wider potentialities to control characteristics of the plasma and the beam as sources of thermal and dynamic effects on the surface of a material being treated.

Unfortunately, it is practically impossible to use conventional arc plasma torches to realise the combined discharge following the above scheme. Coaxial combination of the laser beam with the constricted arc plasma column requires development of special devices, i.e., integrated laser-arc plasma torches, a prototype of which can serve, e.g., the apparatus suggested in [26] and designed for laser + PA welding and cutting. The concept of the integrated laser-arc plasma torches was developed at the E. O. Paton Electric Welding Institute [114–116]*. In addition to exerting a coaxial effect of the laser beam and the arc plasma on a workpiece within the common heating zone, these devices, in fact, will present

---

* Later this concept was taken and patented [117] by the Plasma-Laser Technologies Ltd. (Israel).

a new class of plasma generators acting on the basis of the laser-arc discharge as a special type of the gas discharge. This opens up new prospects for utilising integrated laser-arc plasma torches in the processes that make use of the entire plasma volume, such as surfacing, spraying and plasma chemistry, in addition to the processes of materials joining and treatment associated only with a surface effect on workpiece.

Figure 3.1 shows schematic diagrams of various laser-arc plasma torches having an axially symmetrical design. The integrated plasma torches under consideration are characterised by a peculiar design of the cathode unit that allows the laser beam to be introduced into the arc plasma along the axis of the plasma-shaping channel. For this the cathode unit is made either as a system of pin thermionic cathodes having a circumferential arrangement (Figure 3.1a) or as a tubular refractory cathode operating in the thermionic mode (Figure 3.1b). A material of the thermionic cathode can be tungsten, zirconium, hafnium or other refractory materials, depending upon the kind of a plasma gas. For example, tungsten works well in inert gases (argon, helium, etc.) and is unsuitable for operation in oxidising environments (oxygen, air). Zirconium and hafnium, by contrast, work quite satisfactorily in the air atmosphere, but rapidly fail in the oxygen-free environment, for example in argon. Thermochemical (Zr, Hf) cathodes of the laser-arc plasma torches can be made in the form of one or several inserts pressed-in into a copper casing with a hole made in it for the laser beam to pass through and for part of the plasma gas to be fed to the discharge (Figure 3.1c). Where the arc current has relatively low values (tens and hundreds of amperes), the "cold" cathodes made from materials with a high thermal conductivity, for example, a massive copper cathode with a hole made in it to introduce the laser beam into the plasma and feed the plasma gas (Figure 3.1d), can be used in the integrated laser-arc plasma torches instead of the hot thermionic cathodes. To decrease erosion of such a cathode, it is advisable to use well-known methods for forced displacement of the cathode arc spot over the electrode surface under the effect of gas dynamic or magnetic forces [72].

Similarly to the arc plasma generators, the laser-arc plasma torches can be of direct action, when a workpiece is used as one of the arc electrodes (see Figure 3.1a), or indirect action, when the laser-arc discharge is used for generating the currentless plasma jet (see Figures 3.1b–d). Direct action integrated plasma torches can be used for combined welding, surfacing and cutting of metals, whereas indirect action ones can be used for treatment of dielectric materials, spraying processes, plasmachemical technologies and for research purposes. Note here that the indirect action plasma torch for laser + PJ cutting, welding, hardening and spraying suggested in [23] is not a laser-arc one, since the laser beam there interacts with the currentless plasma jet, rather than with the electric arc column plasma (see Figure 1.9).

As to the methods for spatial stabilisation of the combined discharge in laser-arc plasma torches to ensure stable parameters of the generated plasma and prevent an intensive fracture of the electrodes and plasma-shaping channels, use can be made of the gas stabilisation methods initially developed for arc plasma torches. For example, it is possible to use the vortex stabilisation of the discharge, which involves utilisation of a whirled flow of the plasma gas (see Figures 3.1a, c, d). Whirling of the gas is provided by its tangential feed into the vortex chamber from which the gas, having a torque, flows out to the plasma-shaping channel. The centrifugal forces formed in this case cause formation of a radial gradient of gas static pressure in the discharge with a minimum pressure and, hence, density of the plasma at the channel axis. As a result, the combined discharge with a density lower than that of the surrounding, relatively cold gas, which is caused by high temperature of the plasma, is reliably retained within the near-axis zone of minimum pressure. To select a method for gas stabilisation of the discharge in the laser-arc plasma torches of an axially symmetrical design, one should bear in mind that the effect of the laser beam on the arc plasma causes a relatively high-temperature current-conducting region formed in the latter

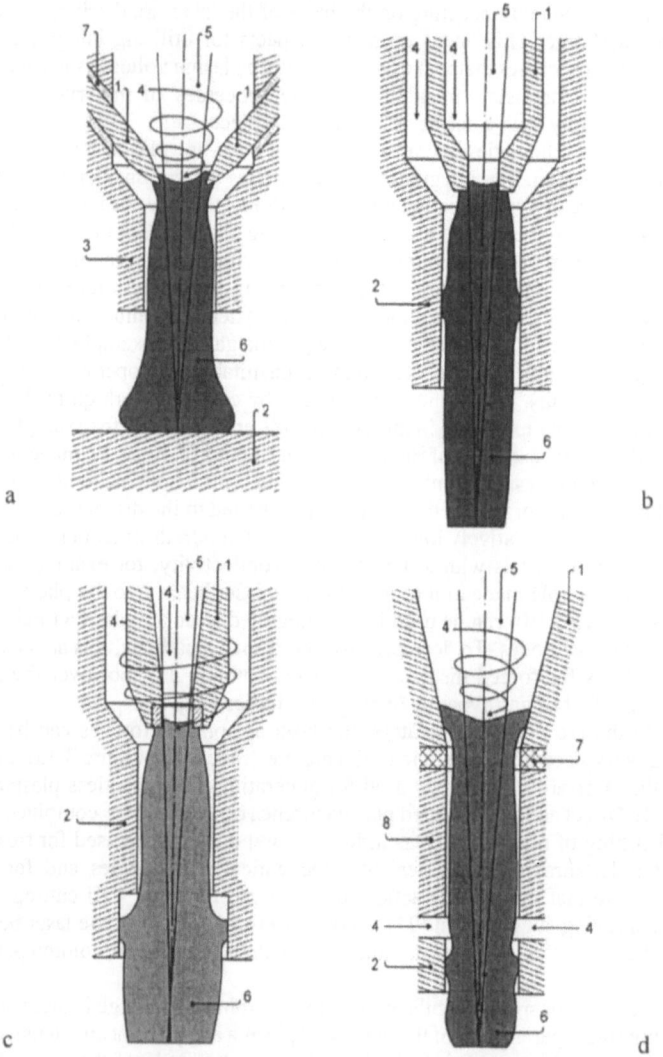

Figure 3.1 Schematic diagrams of integrated laser-arc plasma torches of direct (a) and indirect (b) – (d) action: 1, cathode; 2, anode; 3, plasma-shaping nozzle; 4, plasma-forming gas; 5, focused laser beam; 6, combined discharge plasma; 7, insulator; 8, interelectrode insert.

(see Figures 2.11a, 2.13a), which is strongly connected with the beam axis coinciding with the axis of the plasma-shaping channel. This itself makes the combined discharge spatially more stable than the arc discharge. Therefore, instead of a very efficient vortex stabilisation of the discharge, in integrated plasma torches of the design considered, it could be sufficient to blow over the combined discharge with an axial flow of plasma gas (see Figure 3.1b).

If the cathode unit of the laser-arc plasma torch is made in the form of a tubular thermionic or thermochemical cathode (see Figure 3.1b, c), plasma gas can be fed to the discharge not only through the gap between the cathode and the wall of the plasma-shaping channel, but also through the hole in the cathode. In this case different methods can be used for feeding the internal and external gases, as shown in Figure 3.1c, and their compositions and flow rates can be varied. This allows characteristics of the plasma in the initial region of the flow (see, e.g., Figure 2.17) and, what is particularly important, focusing of the laser beam interacting with it (see Figure 2.27) to be controlled within wide ranges.

Depending upon the feed method, flow rate, composition and pressure of the plasma gas, arc current and parameters of the laser beam, as well as upon design peculiarities, laser-arc plasma torches can be subdivided into torches operating in the laminar mode and those operating in the turbulent mode of flow of the plasma-forming medium. It is a known fact [72, 75] that the electric field intensity in the turbulent flow of the arc plasma is much higher than that in the laminar flow. Therefore, increasing the flow rate of the plasma gas that leads to the development of turbulence is a very efficient method for increasing a specific power input into the plasma at the same arc current. It is apparent that this method of increasing the energy exchange between the discharge and the plasma gas flow can be used also in integrated plasma torches, provided that laser-arc interaction occurs. It should be remembered here that, even in the case of laminar flow of the gas, as was noted at the end of Item 2.3.1, the combined discharge has a higher power density per unit length of the plasma column than the initial arc discharge, as the decrease in the intensity of the field in the plasma caused by the laser radiation effect (see Figure 2.12), hence the decrease in the specific power input of the electric current, are less than the corresponding power addition-ally introduced into the plasma by the laser beam.

Similarly to arc plasma torches, indirect action laser-arc plasma torches can be with a self-adjusting length of the arc, as shown in Figure 3.1b, and with a fixed length of the arc. The latter, in turn, can be subdivided into plasma torches with a gas-dynamic fixation of the anode region (see Figure 3.1c) and plasma torches with an interelectrode insert (see Figure 3.1d). As occurs in conventional arc plasma torches, in laser-arc plasma torches with a smooth cylindrical channel of the nozzle-anode (see Figure 3.1b) a shunting of the arc (break-down of the cold interlayer of gas surrounding the discharge) will lead to variations in its length and, hence, in the parameters of the generated plasma flow, will limit power of the integrated plasma torch and be one of the causes of formation of a drooping volt-ampere characteristic of the combined discharge. To prevent the process of decrease in the arc length, the plasma-shaping channel of the nozzle- anode of the laser-arc plasma torch can be made with a shoulder at its outlet (see Figure 3.1c). The main feature of the functioning of this design of plasma torches is that sudden expansion of the plasma flow results in an intensive mixing of a relatively cold gas near the wall with a heated plasma gas in the discharge, which creates favourable conditions for shunting of the arc. This keeps the average length of the discharge (which is smaller than the self-adjusting one) unchanged, irrespective of the arc current, laser beam power and plasma gas flow rate that are varied within the wide ranges. To have a rising volt-ampere characteristic of the discharge and increase power of the laser-arc plasma torch, its plasma-shaping channel can be elongated using one or several insulated water-cooled metal sections. The interelectrode insert can be insulated from the electrodes by means of dielectric washers, by feeding a cold gas to the gaps between the insert and the electrodes, or by combining both methods (see Figure 3.1d). In the case of using a multisectional interelectrode insert, to increase electrical resistance of this design, the plasma gas can be fed also into the gaps between the sections, which allows the number of the sections to be decreased by increasing their length.

To complete consideration of different arrangements of axially symmetrical laser-arc plasma torches, it should be noted that Figure 3.1, showing schematic diagrams of integrated plasma torches, is far from covering all possible designs of such devices. Also, combinations of elements of laser-arc plasma torches (cathode units and plasma-shaping systems) can be other than those considered above, such as, e.g., in the direct action integrated plasma torch for laser- plasma surfacing having a tubular thermionic cathode. The detailed description of this device, including results of its mathematical modelling and experimental study, is given in Section 3.3.

Serving as hardware to realise various hybrid technological processes, the integrated laser-arc plasma torches will possess a number of important practical advantages over the known arc and optical ones. They will differ from the arc plasma torches, first of all, in extra (due to variation of the laser beam parameters) capabilities for controlling the resulting plasma characteristics. Besides, they will ensure higher spatial and temporal stability of the plasma flow generated (due to laser beam stabilisation of the current-conducting region), this being especially important for a high flow rate of plasma gas, when the development of considerable turbulence is probable. Finally, they will be more reliable, since the above-mentioned stabilisation of the current channel under the effect of laser radiation and decrease in the intensity of the electric field in the arc column, will make it possible to reduce the risk of arc twinning. As compared with optical plasma torches, the laser-arc ones will have (even at low arc currents) a wider range of stable operational conditions with regard to the laser beam parameters and the plasma gas flow rate. The important peculiarity of such plasma torches will be a cardinally new (due to variation of the arc mode parameters) possibility to control the focusing of the laser beam in the combined discharge plasma.

The above advantages of laser-arc plasma torches suggest that the development and practical application of these devices will make it possible not only to improve the known combined processes of welding, cutting and surface heat treatment, but also to create new highly efficient laser-plasma technologies for materials treatment. In particular, the application of such plasma torches for powder deposition and thermal spraying will make it possible to control the processes of heating and acceleration of powder particles by deliberately changing the thermal and gas dynamic characteristics of the generated plasma flow. Besides, it will allow the benefits of the volumetric (laser) [118] and surface (plasma) heating of fine-dispersion particles to be joined. One can expect that the laser-arc plasma generators will find efficient application in various plasmachemical technologies, where a non-equilibrium, non-isothermal plasma is required, which can be easily produced with the help of the combined discharge.

Undoubtedly, the laser-arc plasma torches will be helpful in scientific research as well, particularly, in experimental studies on physics of the gas discharge processes, interaction of the electromagnetic (laser) radiation with the low-temperature plasma and their combined effect on condensed media. While the first two problems have been studied sufficiently well, the problem of the laser-arc action on materials has been studied to a much lesser degree. The interest in this problem has grown since the appearance of the hybrid laser-arc technological processes. It follows from the review of papers dealing with the combined welding, cutting and surface heat treatment, that the qualitative (at the level of estimates) concept of the character of the near-electrode arc processes occurring in the presence of the laser radiation is now available. The peculiarities of the combined laser-plasma interactions with powder materials have not been studied yet applied to such processes as spraying and surfacing. Therefore, the authors of this book hope that research on the laser-arc plasma torches will give a new impetus both to experimental and theoretical investigations of the processes of the combined effect on materials and, eventually, will promote development of new, more efficient hybrid laser-plasma technologies.

3.2 Theoretical Basis and Calculation Technique of Laser-Arc Plasma Torches

Consider the methods for mathematical description of the stationary combined discharge realised by using the integrated plasma torches with an axially symmetrical design, the schematics of which are shown in Figure 3.1. The basic physical phenomena occurring in the case of interaction of the focused laser beam with the arc plasma in such devices are described in Section 2.1. Also, this section suggests the simplest mathematical model of the laser-arc discharge in a channel purged with an axial flow of the plasma gas. This model was developed on the basis of a system of MGD equations in the laminar boundary layer approximation for thermodynamical equilibrium plasma and, as such, does not allow for many peculiarities of the combined discharge associated, for example, with a method of its stabilisation (discharge in a whirled gas flow) and mode of flow of the plasma gas (discharge in a turbulent flow), or for peculiarities of the processes occurring near the electrodes and non-equilibrium of the plasma in a specific laser-arc plasma torch. Therefore, this section is dedicated to description of mathematical models of the laser-arc discharge, which make the most complete allowance for the above peculiarities of its realisation in different integrated plasma torches.

3.2.1 Mathematical model of combined discharge in the plasma torches with laminar gas flow

In a more general case (assuming no applicability of boundary layer approximation), a general system of magnetic-gas dynamic equations (see, e.g., [75]) should be used to describe the thermally equilibrium plasma of the stationary combined discharge in the axially symmetrical laser-arc plasma torches operating in the laminar mode of the plasma-forming gas flow. Allowing for assumptions 1, 3–5 made in Item 2.1.2, this system of equations on cylindrical coordinates, the OZ axis of which coincides with the axis of the plasma-shaping channel of the plasma torch, will have the following form:

$$\rho C_p \left(v \frac{\partial T}{\partial r} + u \frac{\partial T}{\partial z} \right) = \frac{1}{r} \frac{\partial}{\partial r} \left(r \chi \frac{\partial T}{\partial r} \right) + \frac{\partial}{\partial z} \left(\chi \frac{\partial T}{\partial z} \right) + \frac{j_r^2 + j_z^2}{\sigma} + \kappa_\omega \langle S \rangle - \psi; \tag{3.1}$$

$$\rho \left(v \frac{\partial v}{\partial r} + u \frac{\partial v}{\partial z} - \frac{w^2}{r} \right) = - \frac{\partial p}{\partial r} - \mu^0 j_z H_\varphi + \frac{2}{r} \frac{\partial}{\partial r} \left(r \eta \frac{\partial v}{\partial r} \right) -$$

$$- \frac{2}{3} \frac{\partial}{\partial r} \left\{ \eta \left[\frac{1}{r} \frac{\partial (rv)}{\partial r} + \frac{\partial u}{\partial z} \right] \right\} + \frac{\partial}{\partial z} \left[\eta \left(\frac{\partial u}{\partial r} + \frac{\partial v}{\partial z} \right) \right] - 2\eta \frac{v}{r^2};$$

$$\rho \left[v \frac{\partial (rw)}{\partial r} + u \frac{\partial (rw)}{\partial z} \right] = \frac{1}{r} \frac{\partial}{\partial r} \left\{ \eta \left[\frac{\partial (rw)}{\partial r} - 2w \right] \right\} + \frac{\partial}{\partial z} \left[\eta \frac{\partial (rw)}{\partial z} \right];$$

$$\rho \left(v \frac{\partial u}{\partial r} + u \frac{\partial u}{\partial z} \right) = - \frac{\partial p}{\partial z} + \mu^0 j_r H_\varphi + \frac{1}{r} \frac{\partial}{\partial r} \left[r \eta \left(\frac{\partial u}{\partial r} + \frac{\partial v}{\partial z} \right) \right] +$$

$$+ 2 \frac{\partial}{\partial z} \left(\eta \frac{\partial u}{\partial z} \right) - \frac{2}{3} \frac{\partial}{\partial z} \left\{ \eta \left[\frac{1}{r} \frac{\partial (rv)}{\partial r} + \frac{\partial u}{\partial z} \right] \right\}; \tag{3.2}$$

$$\frac{1}{r}\frac{\partial}{\partial r}(r\rho v) + \frac{\partial}{\partial z}(\rho u) = 0, \tag{3.3}$$

where $w\,(r, z)$ is the tangential component of the plasma velocity, the rest of the designations corresponding to those used in Chapter 2. These equations describe the axially symmetrical flow of the plasma generated in the laser-arc plasma torches with whirling of the plasma gas. To describe the combined discharge plasma in the integrated plasma torches with an axial gas flow, equations of motion (3.2) can be simplified to a considerable degree by assuming in them that $w = 0$.

The system of equations (3.1) to (3.3) should be supplemented with Maxwell's equations:

$$\frac{\partial E_r}{\partial z} - \frac{\partial E_z}{\partial r} = 0, \tag{3.4}$$

$$\frac{1}{r}\frac{\partial (rH_\phi)}{\partial r} = j_z; \quad \frac{\partial H_\phi}{\partial z} = -j_r, \tag{3.5}$$

where $E_r\,(r, z)$ and $E_z\,(r, z)$ are the radial and axial components of the intensity of the electric field in the plasma; $H_\phi\,(r, z)$ is the azimuthal component of the magnetic field of the discharge current, and supplemented with the Ohm's law:

$$j_r = \sigma E_r; \quad j_z = \sigma E_z, \tag{3.6}$$

where $j_r\,(r, z)$, $j_z\,(r, z)$ are the radial and axial components of the current density in the plasma, respectively. In addition, to solve the system of equations (3.1) through (3.5), it is necessary to use integrated conditions of conservation of the plasma gas flow rate within the plasma-shaping channel:

$$G = 2\pi \int_0^{R_C} \rho u r\, dr, \tag{3.7}$$

as well as the electric current in the discharge:

$$I = 2\pi \int_0^{R_\sigma} j_z r\, dr, \tag{3.8}$$

where: G is the total mass flow rate of the plasma gas, which can be varied along the length of the channel due to a lateral blowing in of the gas, as shown, e.g., in Figure 3.1d; $R_C\,(z)$ is the radius of the plasma-shaping channel of the plasma torch; I is the discharge current and $R_\sigma\,(z)$ is the radius of the current-conducting region. In the case of the indirect action laser-arc plasma torches (see Figures 3.1b–d), the electric current determined by means of the integral relationship (3.8) should decrease within the region of fixation of the arc to the nozzle-anode from a constant value in the electric current zone of the plasma flow to zero in the currentless plasma. It should be noted here that the system of equations (3.1) to (3.3) can be used also to describe the currentless plasma jet generated by indirect action integrated plasma torches. As the only mechanism of heating the plasma in the currentless zone of the flow is the absorption of the laser radiation energy and there are no electromagnetic forces to accelerate the plasma, equations (3.1) and (3.2) can be simplified by assuming that $j_r = j_z = 0$.

The $\kappa_\omega \langle S \rangle$ term in equation (3.1) describes an additional release of the energy in the arc plasma at its heating by the laser. Besides, as the plasma system is assumed to be cylindrically symmetrical and distribution of the radiation intensity across the section of the initial beam in a general case can have no cylindrical symmetry (as, e.g., is the case of the high-order Gaussian modes [80]), to calculate the energy of laser radiation absorbed by the plasma it is possible to use the angle-averaged distribution of its intensity $\langle S \rangle$ (r, z) [115]:

$$\langle S \rangle = \frac{1}{2\pi} \int_0^{2\pi} S d\varphi .$$ (3.9)

Here $S(r, \varphi, z)$ is the true distribution of radiation intensity in the laser beam with allowance for refraction and absorption in the plasma, which is found using the quasi-optical approximation, as was done in Item 2.1.2. Assuming that vector of the electric field of the beam has the form of (2.11), where the complex amplitude of the field A_ω is a function of r, z and φ, the parabolic equation to find it in the selected cylindrical system of coordinates with the OZ axis directed towards propagation of the initial laser beam can be written in the following form [115]:

$$- 2ik \frac{\partial A_\omega}{\partial z} = \frac{1}{r} \frac{\partial}{\partial r} \left(r \frac{\partial A_\omega}{\partial r} \right) + \frac{1}{r^2} \frac{\partial^2 A_\omega}{\partial \varphi^2} + k^2 (\varepsilon_\omega - 1) A_\omega ,$$ (3.10)

where $\varepsilon_\omega (r, z)$ is the spatial distribution of the complex dielectric permittivity of the combined discharge plasma, which is assumed to be axially symmetrical. The value S to be determined, which in fact is an axial component of density of the flow of the electromagnetic energy of the laser beam, can be calculated from the found distribution of the complex amplitude of the electric field of the beam using relationship (2.14).

To write down the energy equation (3.1) for the laser-arc discharge plasma in a direct action integrated plasma torches (see Figure 3.1a), along with heating the plasma by the initial laser beam, one should also take into account, generally speaking, its heating by laser radiation which is reflected from the surface of the anode (workpiece). If the reflecting surface is normal to the plasma torch axis, the incident and reflected beams can also be represented as quasi-plane waves (2.11) propagating in opposite directions and having complex amplitudes $A_{1\omega}(r, \varphi, z)$ and $A_{2\omega}(r, \varphi, z)$, respectively. In this case within the frames of the quasi-optical approximation the $A_{1\omega}$ value will satisfy conditions of equation (3.10), and the $A_{2\omega}$ value—conditions of the equation obtained from (3.10) by substituting $-k$ for k:

$$2ik \frac{\partial A_{2\omega}}{\partial z} = \frac{1}{r} \frac{\partial}{\partial r} \left(r \frac{\partial A_{2\omega}}{\partial r} \right) + \frac{1}{r^2} \frac{\partial^2 A_{2\omega}}{\partial \varphi^2} + k^2 (\varepsilon_\omega - 1) A_{2\omega}.$$ (3.11)

It is not difficult to show that also in the case under consideration the term that describes laser heating of the plasma in equation (3.1) can be represented as $\kappa_\omega \langle S \rangle$, where $\langle S \rangle$, allowing for (3.9), implies:

$$\langle S \rangle = \frac{1}{4\pi} \sqrt{\varepsilon^0 / \mu^0} \int_0^{2\pi} (|A_{1\omega}|^2 + |A_{2\omega}|^2) \, d\varphi .$$ (3.12)

Therefore, we have a closed system of equations which make the basis of the local thermal equilibrium (LTE) model of the stationary laser-arc discharge in the axially symmetrical integrated plasma torches (see Figure 3.1) operating in the laminar mode of plasma gas flow. This system should be supplemented with relationships (2.10) and (2.13) to determine thermodynamic parameters, transport coefficients and optical properties of the thermal equilibrium plasma as functions of its temperature, pressure and composition of the plasma-forming medium (see, e.g., Section 2.2).

Possibilities of finding an analytical solution to the system of non-linear differential equations (3.1) through (3.5), (3.10) and (3.11) are very limited. Therefore, it is necessary to use numerical methods to solve it. In doing this, instead of the continuity equation (3.3) it is more convenient to use an equation for pressure, which can be obtained by taking divergence from a vector momentum equation (3.2) and using, instead of two Maxwell's equation, one equation of the second order for the magnetic field intensity, which is obtained by excluding the electric field from the above equations by means of relationships (3.6). As a result, the initial system of the equations is converted into a system of six differential equations of the second order with regard to magnetic-gas dynamic variables: T, v, w, u, p and H_φ, as well as two parabolic equations with regard to complex amplitudes of the laser radiation field: $A_{1\omega}$ and $A_{2\omega}$. To solve this system of the equations which describe laser-arc discharge in the integrated plasma torches under consideration, it is necessary to assign appropriate boundary and input conditions, which depend upon the type and design of the particular plasma torch, mode of a laser beam used, method of stabilisation of the discharge and other peculiarities associated with its realisation.

Choice of boundary conditions at the axis of the discharge is based on consideration of the axial symmetry of the problem, so that at $r = 0$; $z > 0$ they are as follows:

$$\frac{\partial T}{\partial r} = 0; \quad v = 0; \quad w = 0; \quad \frac{\partial u}{\partial r} = 0; \quad \frac{\partial p}{\partial r} = 0; \quad H_\varphi = 0; \quad \frac{\partial A_{(1,2)\,\omega}}{\partial r} = 0. \tag{3.13}$$

It is assumed that at an external boundary of the discharge within the limits of the plasma-shaping channel of the plasma torch the temperature of the plasma gas is equal to that of a cooled wall of the channel T_C, and that conditions of impenetrability of the wall and "sticking" of the gas are satisfied. Also, it is assumed that there is no electric current to the wall and that the transverse size of the laser beams (both initial and reflected) is substantially smaller than the diameter of the channel. Therefore, the selected boundary conditions at $r = R_C$; $z < L_C$, where L_C is the length of the plasma-shaping channel, are as follows:

$$T = T_C; \quad v = 0; \quad w = 0; \quad u = 0; \quad p = p_C; \quad H_\varphi = \frac{I}{2\pi R_C}; \quad A_{(1,2)\,\omega} = 0. \tag{3.14}$$

Here p_C (z) is the gas pressure at the wall, which can be determined by using an integral equation (3.7). In the case of the laser-arc plasma torch with a gas dynamic fixation of the anode region of the discharge (see Figure 3.1c), the boundary conditions should be assigned also for $z = Z_s$; R_C $(Z_s - 0) < r < R_C$ $(Z_s + 0)$ (Z_s is the plane location of a sudden expansion of the plasma-shaping channel). Here conditions for temperature, velocity and pressure of the gas are assigned similarly to those in (3.14) and distribution of the magnetic field intensity is determined from the following expression:

$$H_\varphi = \frac{I}{2\pi r}.$$

To consider the plasma flow in laser-arc plasma torch with a lateral blowing in of a cold plasma gas (see, e.g., Figure 3.1d), at $r = R_C$; $Z_{1g} < z < Z_{2g}$ (Z_{1g} and Z_{2g} are the boundaries of a region where this gas is fed) it is necessary to use conditions of $v = v_C$ and $w = w_C$ instead of the second and third conditions of (3.14) and employ expression:

$$G = G_0 + 2\pi R_C \rho_C \int_{Z_{1g}}^{z} v_C \, dz$$

to calculate the G value in this region. Here v_C and w_C are respectively the radial and tangential ($w_C = 0$, unless the gas is whirled) components of the velocity of the gas fed, which has a temperature at the wall equal to T_C; ρ_C is its density at this temperature and G_0 is the mass flow rate of the plasma gas at the initial cross-section of the channel.

To consider the discharge in indirect action integrated plasma torches, where the plasma-shaping nozzle at the same time serves as the anode of the arc (see Figure 3.1b–d), in the region of the anode fixation of the discharge, i.e., at $r = R_C$; $Z_{1a} < z < Z_{2a}$ (Z_{1a} and Z_{2a} are the boundaries of the said region), it is necessary to use conditions:

$$T = T_a; \; H_\varphi = \frac{I}{2\pi R_C} - \int_{Z_{1a}}^{z} j_a \, dz \, ,$$

instead of the first and sixth conditions of (3.14). Here T_a is the temperature of the plasma in the anode region and j_a is the current density at the anode, which are assigned from an experiment or calculated on the basis of available models of the anode processes of the arc discharge [119, 120].

It is assumed that conditions of smooth contingency with the environment exist at the external boundary of an open region of the discharge. For example, the assigned boundary conditions at $r = R$; $z > L_C$, where R is the radius of a calculation domain in the open region of the flow, are as follows:

$$T = T_{ext}; \; v = 0; \; w = 0; \; u = u_{ext}; \; p = p_{ext}; \; H_\varphi = \frac{I}{2\pi R}. \tag{3.15}$$

Here: u_{ext} is the velocity of the axial flow of the gas that blows over the discharge ($u_{ext} = 0$, if the plasma flows to the quiescent environment); T_{ext} and p_{ext} are its temperature and pressure, while the value of R is selected so high that at $r = R$ all characteristics of the flow meet requirements of the type of (2.19). As the R value thus determined is as a rule larger than R_C (L_C), also it is necessary to assign boundary conditions for $z = L_C$; R_C (L_C) $< r < R$, i.e., in the plane of an outlet section of the plasma torch. These conditions for gas dynamic variables are selected to be similar to (3.14) at the nozzle surface and to (3.15) in the region where the gas to blow over the discharge is fed, while for the magnetic field the relationship of $H_\varphi = I/(2\pi r)$ is used.

As far as the boundary conditions for a complex amplitude of the electric field of the beam in the open region of the discharge are concerned, supposing that transverse size of the region occupied by the laser radiation field is substantially smaller than the diameter of the plasma-shaping channel, also at $z > L_C$ the value of, e.g., R_C can be selected as the radius of the calculation domain for equations (3.10), (3.11) and conditions of (3.14) can be used for $A_{(1,2)\,\omega}$ at all values of z. In addition, to solve equations (3.10) and (3.11) which are three-dimensional, it is necessary to assign cyclic boundary conditions by φ, i.e., this requires that the following condition be met at any values of r and z:

$$A_{(1,2)\,\omega}(r, \varphi, z) = A_{(1,2)\,\omega}(r, \varphi + 2\pi, z). \tag{3.16}$$

Boundary conditions at the inlet section of the plasma-shaping channel, i.e., at $z = 0$ and $r < R_C$ (0), are usually assigned in the form of the following relationships:

$$T = T(r, 0); \; v = v(r, 0); \; w = w(r, 0); \; u = u(r, 0);$$

$$\frac{\partial p}{\partial z} = \frac{\partial p}{\partial z}(r, 0); \; H_\varphi = H_\varphi(r, 0); \; A_{1\omega} = A_{1\omega}(r, \varphi, 0). \tag{3.17}$$

As the inlet section of the channel is selected, as a rule, to be located directly under the cathode (see Figure 3.1a–c), the explicit form of these relationships, particularly the radial distributions of the plasma temperature, current density and, hence, intensity of the magnetic field near the cathode, should be determined by taking into account the cathode processes that occur in a specific laser-arc plasma torch. As far as the integrated plasma torch with a "cold" copper cathode (see Figure 3.1d) is concerned, here the inlet section should be selected to be above the region of the cathode fixation of the discharge. In this case the assignment of boundary conditions at $z = 0$; $r < R_C(0)$ can be substantially simplified by assuming that $T(r, 0) = T_C$ and $H_\varphi(r, 0) = 0$. But in doing so, at $r = R_C$; $Z_{1c} < z < Z_{2c}$ (Z_{1c} and Z_{2c} are the boundaries of the cathode fixation region) it will be necessary to use the following conditions:

$$T = T_c; \quad H_\varphi = \int_{Z_{1c}}^{z} j_c dz \,,$$

where: T_c is the plasma temperature in the cathode region and j_c is the current density at the cathode, instead of the first and sixth conditions of (3.14). Therefore, in this case it is also impossible to correctly assign boundary conditions without a detailed consideration of the cathode processes.

To find distributions of the velocity and pressure gradient of the plasma gas at the inlet section of the channel, it is possible to use simplified, one-dimensional gas dynamic equations (see, e.g., [75]) by taking into account the integral relationships of conservation of the mass flow rate of the gas and its whirling moment in plasma torches with a vortex stabilisation of the discharge. For example, in cases shown in Figure 3.1a, d the distribution of the flow rate of the plasma gas in the said section should satisfy the following integral relationships:

$$G_0 = 2\pi \int_0^{R_C(0)} \rho u r dr; \quad M_0 = 2\pi \int_0^{R_C(0)} \rho w r^2 dr \,, \tag{3.18}$$

where M_0 is the whirling moment of the gas at the inlet section of the calculation domain. If the cathode unit of the laser-arc plasma torch is made in the form of a tubular thermionic or thermochemical cathode (see Figure 3.1b, c), relationships (3.18) should be written as follows:

$$G_0 = G_1 + G_2; \quad G_1 = 2\pi \int_0^{R_1} \rho u r dr; \quad G_2 = 2\pi \int_{R_2}^{R_C(0)} \rho u r dr \,, \tag{3.19}$$

where: G_1 is the flow rate of the plasma gas fed through a hole in the cathode; R_1 is the radius of this hole; G_2 is the flow rate of the gas fed through the gap between the cathode and a wall of the plasma-shaping channel; R_2 is the external radius of a tip of the cathode, and:

$$M_1 = 2\pi \int_0^{R_1} \rho w r^2 dr; \quad M_2 = 2\pi \int_{R_2}^{R_C(0)} \rho w r^2 dr \,, \tag{3.20}$$

where: M_1 and M_2 are the whirling moments of the said gases, in the cases shown in Figures 3.1b ($M_1 = M_2 = 0$) and c ($M_1 = 0, M_2 \neq 0$).

Radial distribution of a complex amplitude of the electric field of the initial laser beam in the inlet plane of the channel, which is an input condition for solving parabolic equation (3.10), is determined by the mode and power of a beam used. For example, for the Gaussian beams (TEM_{mn} modes) the above distribution is assigned by relationships [80]:

$$A_{1\omega} = A_{1\omega F} \frac{r_F}{r_z} H_m \left(\sqrt{2} \frac{r \cos\varphi}{r_z} \right) H_n \left(\sqrt{2} \frac{r \sin\varphi}{r_z} \right) \times$$

$$\times \exp \left\{ -\frac{r^2}{r_2^2} + i \left[k \frac{r^2}{2R_z} - (m + n + 1) \varphi_z \right] \right\}.$$

$$(3.21)$$

Here: r_F is the initial beam radius at the beam waist plane ($z = F$); H_m and H_n are Hermite polynomials of degree m and n (m, $n = 0, 1, 2, 3,...$); r_z (z), R_z (z) and φ_z (z) are defined by expressions (2.28) to (2.31); $A_{1\omega F}$ is found from the relationship for the total beam power with allowance for (2.14) giving:

$$A_{1\omega F} = \sqrt{(4Q^0 / 2^{m+n} \, m! n! \pi r_F^2)} \, (\mu^0 / \varepsilon^0)^{1/2} \, , \qquad (3.22)$$

where Q^0 is the power of the initial laser beam.

As the equations for T, v, w, u, p and H_φ are the equations of the second order, it is impossible to solve them without the assignment of boundary conditions in the plane of $z = L$, where L is the length of the calculation domain of the discharge. To consider the laser-arc discharge in the direct action integrated plasma torches (see Figure 3.1a), this plane should be selected so that it is located in the immediate vicinity of the anode (workpiece) surface. Then, assuming the anode to be non-evaporating, the boundary conditions at $z = L$ and $r < R$ can be written, e.g., in the following form:

$$T = T(r, L); \; v = 0; \; w = 0; \; u = 0; \; \frac{\partial p}{\partial z} = \frac{\partial p}{\partial z}(r, L); \; H_\varphi = H_\varphi(r, L) . \qquad (3.23)$$

The explicit form of radial distributions of the intensity of the magnetic field and distribution of the pressure gradient in the direction normal to the selected surface can be easily determined from the first equation in (3.5) and the third equation in (3.2), respectively, by assuming in them that $j_r = j_{ra}(r), j_z = j_{za}(r)$ (j_{ra} and j_{za} are the radial and axial components of the current density in the anode region) and taking into account that the velocity of the plasma near the anode surface is equal to zero. Therefore, the correct assignment of boundary conditions (3.23) is reduced to finding a distribution of the plasma temperature and current density in the anode region. In turn, this problem can not be solved without a detailed consideration of the anode processes occurring in the laser-arc discharge, i.e., the processes that take place at the combined interaction of plasma and laser radiation with the anode surface.

In addition to conditions of (3.23), at $z = L$ it is necessary to determine also the radial distribution of the complex amplitude of the field of the laser beam reflected from the anode surface, which will be an input condition for parabolic equation (3.11). To find this condition, it is possible to use a formula that relates fields of the incident and reflected beams at the surface of metal that is assumed to homogeneous and isotropic [121, 122], which can be written as follows at a normal incidence of the laser beam:

$$A_{2\omega}(r, \varphi, L) = \frac{r_{s\omega} - 1}{r_{s\omega} + 1} A_{1\omega}(r, \varphi, L) , \qquad (3.24)$$

where $r_{s\omega}$ is the surface impedance of metal at the laser radiation frequency [81, 121].

In calculation of characteristics of the combined discharge plasma generated in the indirect action laser-arc plasma torches (see Figure 3.1b–d) the assignment of boundary conditions in the inlet plane of the calculated region for a currentless plasma jet can be simplified to a certain extent. As the system of equations (3.1) through (3.3) used to describe plasma in the currentless region of the flow contains no electrical or magnetic values*, there is no need to assign the last condition of (3.23) for the magnetic field intensity, hence there is no need to find distribution of the current density in the anode region either. Nevertheless, if the $z = L$ plane coincides with the surface of a workpiece, for the correct assignment of the radial distribution of the plasma temperature $T(r, L)$, it is necessary to use a self-consistent consideration of physical processes occurring at the currentless plasma—condensed matter interface in the presence of laser radiation. As far as the distribution of an amplitude of the field of laser radiation reflected from the workpiece surface $A_{2\omega}(r, \varphi, L)$, is concerned, it is still possible to use relationship (3.24) to determine it.

Finally, if the plasma jet flowing out to a free space is considered, or if it is supposed that the workpiece surface is located at a sufficiently large distance from the outlet section of the plasma torch and outside the calculation domain, at $z = L, r < R$ the "soft" boundary conditions [75] can be used instead of conditions of (3.23):

$$\frac{\partial^2 \phi}{\partial z^2} = 0 , \tag{3.25}$$

where $\phi = \{T, v, w, u, p\}$. Besides, input condition (3.24) becomes unnecessary, as in this case the contribution to the energy balance of the plasma by the reflected laser radiation can be ignored (see Item 2.1.1) and, hence, equation (3.11) can be left unsolved.

3.2.2 Laser-arc discharge in the plasma torches with turbulent gas flow

The high-power laser-arc plasma torches intended, for example, for hybrid cutting or spraying, which are designed for large amounts of the plasma gas passing through them, are characterised, as a rule, by a turbulent mode of plasma flow. In this case parameters of plasma flow vary in an arbitrary manner in time and space about their mean values, and the fundamental effect on the character of turbulence may be exerted by such factors as the natural magnetic field of the discharge, Joule heat and absorption of the energy of laser radiation by the plasma. The Reynolds's approach is usually used to describe such flows. This approach consists in the resolution of the independent variables, which are part of the conservation equations into the time-averaged values and pulsating components, followed by time averaging of the resulting equations.

Consider the simplest model of stationary laser-arc discharge in axially symmetrical integrated plasma torches (see Figure 3.1) operating in a turbulent mode of the plasma gas flow, where the effect of factors of the electromagnetic nature on the character of turbulence can be ignored, i.e., where turbulence can be considered to be purely gas-dynamic. The latter implies that pulsations of the current density, magnetic field, intensity of laser radiation, electrical conductivity and absorptivity can be ignored and these values are regarded as averaged. Equations that describe this turbulent flow in the case of the thermal equilibrium laser-arc plasma can be derived from a general system of MGD equations, which have the form of (3.1) to (3.3) in the cylindrical system of coordinates. For this purpose, according to [123], write the thermophysical and gas-dynamic variables** included into the above equations in the following form:

---

* Except for the $\kappa_\omega \langle S \rangle$ term in equation (3.1), which describes the laser radiation heating of the plasma.
** Normally, pulsations of coefficients C_p, χ and η are small, therefore, further on they will be ignored.

$$\rho = \bar{\rho} + \rho'; \ p = \bar{p} + p';$$
$$T = \bar{T} + T''; \ v = \tilde{v} + v''; \ w = \tilde{w} + w''; \ u = \tilde{u} + u'', \tag{3.26}$$

where:

$$\bar{f} = \frac{1}{\Delta t} \int\limits_{t}^{t+\Delta t} f \, dt$$

is the averaging according to Reynolds;

$$\tilde{g} = \frac{\overline{\rho g}}{\rho}$$

is the averaging with the density used as the weighting function (averaging according to Favre [124]), which is employed for the description of turbulent gas flows with variable density; f', g'' are the corresponding pulsations which meet the following apparent relationships: $\bar{f}' = 0$, $\overline{\rho g''} = 0$. Substitute expressions (3.26) in the initial system and do the time averaging of the resulting equations by assuming the ternary correlations of the pulsations to be small compared with the binary correlations. Using the Bossinesq hypothesis, according to which the Reynolds stresses and heat flows containing the velocity and temperature pulsations can be related to the mean deformation tensor and the mean temperature gradient through the scalar coefficients of the apparent eddy viscosity and thermal conductivity, it can be shown that the system of equations for the averaged values of $\bar{p}$, $\bar{T}$, $\tilde{v}$ and $\tilde{u}$ will preserve the form of (3.1) to (3.3), if the following expressions are used instead of η and χ:

$$\bar{\eta} = \eta + \eta_t; \ \bar{\chi} = \chi + \chi_t. \tag{3.27}$$

Here: η and χ are the coefficients of molecular viscosity and thermal conductivity, which can be determined, for example, from formulae (2.46) and (2.47); η_t and χ_t are the coefficients of apparent eddy viscosity and thermal conductivity, which are estimated from simple algebraic models for the engineering calculations.

Before we start describing models of eddy viscosity and thermal conductivity, let us write down the said system of equations for the averaged values in the axially symmetrical boundary layer approximation. With large flow rates of the plasma gas, i.e. particularly in the cases where the turbulent mode of the plasma flow is realised, this approximation suits the description of the laser-arc discharge in the integrated plasma torches under consideration (see Figure 3.1), as the characteristics of the flow in this case meet the following conditions:

$$|\tilde{v}| << |\tilde{w}|, \ |\tilde{u}|; \ \left|\frac{\partial \tilde{\phi}}{\partial z}\right| << \left|\frac{\partial \tilde{\phi}}{\partial r}\right|,$$

where $\tilde{\phi} = \{ \bar{T}, \tilde{v}, \tilde{w}, \tilde{u} \}$. Using a standard procedure for deriving the boundary layer equations [125], the resulting system of equations for the time averaged characteristics of the combined discharge plasma in a turbulent gas flow will be as follows:

$$\bar{\rho} C_p \left(\tilde{v} \frac{\partial \bar{T}}{\partial r} + \tilde{u} \frac{\partial \bar{T}}{\partial z} \right) = \frac{1}{r} \frac{\partial}{\partial r} \left(r \bar{\chi} \frac{\partial \bar{T}}{\partial r} \right) + \frac{j_r^2 + j_z^2}{\sigma} + \kappa_\omega \langle S \rangle - \psi ; \tag{3.28}$$

$$\bar{\rho} \left[\tilde{v} \frac{\partial (r\bar{w})}{\partial r} + \tilde{u} \frac{\partial (r\bar{w})}{\partial z} \right] = \frac{1}{r} \frac{\partial}{\partial r} \left\{ r \bar{\eta} \left[\frac{\partial (r\bar{w})}{\partial r} \right] \right\} ;$$

$$\bar{\rho}\left(\tilde{v}\frac{\partial\bar{u}}{\partial r}+\tilde{u}\frac{\partial\bar{u}}{\partial z}\right)=\frac{1}{r}\frac{\partial}{\partial r}\left(r\bar{\eta}\frac{\partial\bar{u}}{\partial r}\right)-\frac{\partial\bar{p}}{\partial z}+\mu^0\bar{j}_r\,H_\varphi\;;$$

$$(3.29)$$

$$\frac{1}{r}\frac{\partial}{\partial r}\,(r\bar{\rho}\tilde{v})+\frac{\partial}{\partial z}\,(\bar{\rho}\,\tilde{u})=0\;.$$

$$(3.30)$$

Here $\tilde{v}=\overline{(\rho v+\rho' v')}/\bar{\rho}$, whereas the difference between the methods of averaging the T, w and u values in the approximation considered disappears, i.e.: $\overline{T}=\tilde{T}$, $\tilde{w}=\overline{w}$ and $\tilde{u}=\overline{u}$ [123]. Equations (3.28) to (3.30) for the turbulent boundary layer describe the gas-dynamic turbulence of the plasma generated in the axially symmetrical laser-arc plasma torches with whirling the plasma gas (see Figure 3.1a, c, d). In this case, the distribution of the averaged pressure $\bar{\rho}$ (r, z) within the plasma-shaping channel $(z < L_C)$ can be calculated from the following formula:

$$\bar{p}=p_{\text{ext}}-\int_z^{L_C}\frac{d\bar{p}_c}{dz}\,dz+\mu^0\int_r^{R_C}j_z\,H_\varphi\,dr-\int_r^{R_C}\bar{\rho}\frac{\overline{w}^2}{r}\,dr\;,$$

$$(3.31)$$

where $d\bar{p}_c/dz\,(z)$ is the gradient of the gas-static pressure on the channel wall, which is determined from the integral condition of conservation of a full plasma gas flow rate:

$$G=2\pi\int_0^{R_C}\bar{\rho}\,\bar{u}\,r\,dr\;.$$

In the open region of the flow $(z > L_C)$ the distribution of pressure is of the following form:

$$\bar{p}=p_{\text{ext}}+\mu^0\int_r^{R}j_z\,H_\varphi\,dr-\int_r^{R}\bar{\rho}\frac{\overline{w}^2}{r}\,dr\;.$$

$$(3.32)$$

To consider turbulent flow of laser-arc discharge plasma in integrated plasma torches with axial feed of the plasma gas (see Figure 3.1b), the equations of motion (3.29) and the expressions of pressure (3.31) and (3.32) can be substantially simplified by assuming in them that $\overline{w}=0$. The boundary layer equations thus obtained for the calculation of averaged characteristics of turbulent flow of combined discharge plasma blown over by axial gas flow, coincide in form with (2.1) to (2.3), (2.6) and (2.9), if expressions (3.27) are used instead of η and χ in (2.1) and (2.2) [71].

The system of equations (3.28) through (3.32) should be supplemented with equations for the determination of averaged values of the magnetic field intensity and current density in the plasma. As the assumptions made in Section 2.1, that the radial component of the current density is small compared with the axial component and that E_z is constant across the discharge section, are known to be absolutely unsuitable for considering the discharge in indirect action laser-arc plasma torches (see, e.g., Figures 3.1b–d) to find the magnetic field, in the general case it is necessary to use the following equation:

$$\frac{\partial}{\partial r}\left[\frac{1}{r\sigma}\frac{\partial\,(rH_\varphi)}{\partial r}\right]+\frac{\partial}{\partial z}\left(\frac{1}{\sigma}\frac{\partial H_\varphi}{\partial z}\right)=0\;,$$

$$(3.33)$$

along with the integral condition of conservation of the electric current (see Item 3.2.1), while to determine j_r and j_z it is necessary to use equations (3.5). It should be noted here that the above system of equations remains valid also for the currentless region of turbulent plasma flow, if $j_r = j_z = 0$ is used in equations (3.28), (3.29) and (3.31), (3.32).

Applicability of the above combined discharge model in the axially symmetrical boundary layer approximation is limited to that region of the flow which is far from a workpiece, as the axial gradients of characteristics of the plasma and its radial velocity become more and more substantial with the distance to its surface. Therefore, for modelling, e.g., the discharge in direct action laser-arc plasma torches (see Figure 3.1a) operating in the turbulent mode of plasma gas flow, it is necessary to use the full system of equations for the averaged values or limit the length of the calculated region for equations (3.28) through (3.33), so that a workpiece (anode) can be considered to be at a sufficient distance from it. In addition, when using the turbulent boundary layer approximation, the latter assumption makes it possible to ignore the laser radiation reflected from the anode surface. Therefore, to determine the laser radiation intensity $\langle S \rangle$ which is included in the equation of energy (3.28), it is sufficient to solve only equation (3.10) and make allowance for relationships (2.14) and (3.9).

The system of equations (3.28) through (3.33) and (3.10) should be supplemented with relationships (2.10) and (2.13), which determine the averaged density, specific heat, molecular transport coefficients and optical properties of the plasma as a function of the averaged values of temperature and pressure. In addition, as the resulting coefficients of viscosity and thermal conductivity (3.27) contain the turbulent components, it is also necessary to establish the relationship of η_t and χ_t with the averaged characteristics of the plasma flow, i.e., select the model of turbulence.

To determine the turbulent viscosity coefficient of plasma within a plasma-shaping channel of the laser-arc plasma torch, it is possible to use, for example, a combined model of turbulence which, as is shown in [75], gives the best agreement with the experiment for the entire set of characteristics of the electric-arc flows in the channels:

1 central region (Prandtl model [126])

$$\eta_t = \overline{\rho} l_m^2 \sqrt{(\partial \overline{w} / \partial r)^2 + (\partial \overline{u} / \partial r)^2} \, , \tag{3.34}$$

where $l_m = 0.41 (R_C - r)$ is the mixing length;

2 near-wall region (Deissler model [127])

$$\frac{\eta_t}{\eta} = 0.0154 \, u^+ \, Y^+ \, [1 - \exp (- \, 0.0154 \, u^+ \, Y^+)] \, ;$$

$$u^+ = \frac{\sqrt{\overline{w}^2 + \overline{u}^2}}{v*}; \quad Y^+ = \frac{\overline{\rho} \, (R_C - r) \, v*}{\eta} \, , \tag{3.35}$$

where:

$$v* = \sqrt{\tau_C / \rho_C}$$

is the dynamic velocity;

$$\tau_C = \eta \sqrt{(\partial \overline{w} / \partial r)^2 + (\partial \overline{u} / \partial r)^2} \, , \; at \; r = R_C$$

is the tangential tension and $\overline{\rho}_C$ is the averaged plasma density at the channel wall. The range of application of relationships (3.35) is determined by the $0 < Y^+ < 26$ condition. At $Y^+ > 26$, η_t can be calculated from formula (3.34).

For the open region of the laser-arc plasma flow ($z > L_C$) it is possible to use one of the models of turbulence of a round coaxial jet [123], for example, the Madni-Pletcher model

[128], which, in the case of a variable-density gas flow, is described by the following expression:

$$\eta_t = \overline{\rho} \frac{0.03 \, (1 + 2.13 \, g^{+\,2})}{R_C} \int_r^R |\overline{u} - u_{\text{ext}}| r \, dr \,, \tag{3.36}$$

where:

$$g^+ = \frac{\pi R_C^2 \, \overline{\rho} \, \overline{u}}{G} \,;$$

R_C and G are the radius of the plasma-shaping channel and the full mass flow rate of the plasma gas at the exit section of the plasma torch; u_{ext} is the non-disturbed velocity of a gas flow that blows over the plasma. It should be emphasised here that if a turbulent plasma jet enters an environment whose properties differ from those of the plasma gas, in the calculation of thermodynamic parameters, transport coefficients and optical properties of the plasma in the open region of the flow, it is necessary, generally speaking, to take into account the admixture of the surrounding gas to the plasma gas. For instance, by expressing the local value of the relative concentration of the surrounding gas in the mixture in terms of $\delta \, (r, z)$, the averaged value of its density $\overline{\rho}_m$ can be found from the following relationship:

$$\overline{\rho}_m = (1 - \delta) \, \overline{\rho} + \delta \, \overline{\rho}_{\text{ext}}, \tag{3.37}$$

where: $\overline{\rho} = \rho \, (\overline{T}, \, p_{\text{ext}})$ is the density of the plasma-forming medium; $\overline{\rho}_{\text{ext}} = \rho_{\text{ext}} \, (\overline{T}, \, p_{\text{ext}})$ is the density of the surrounding gas and p_{ext} is its pressure. Then, using the condition of similarity of the excessive temperature and concentration profiles for a free turbulent jet [125] expression (3.37) will yield:

$$\delta = 1 - (\overline{T} - T_{\text{ext}}) \, G / 2\pi \int_0^R (\overline{T} - T_{\text{ext}}) \, \overline{\rho} \, (\overline{u} - u_{\text{ext}}) \, r \, dr \,, \tag{3.38}$$

where T_{ext} is the non-disturbed temperature of the surrounding. The value of δ thus obtained can be used to calculate the rest of the parameters of the plasma mixture (see, e.g., Section 2.2).

As far as the coefficient of turbulent thermal conductivity of the plasma χ_t is concerned, for flow within the channel of the integrated plasma torch and for that in the open region, it can be expressed in terms of the turbulent viscosity coefficient of the plasma for the corresponding region using the following relationship:

$$\chi_t = \eta_t \frac{C_p}{\text{Pr}_t} \,, \tag{3.39}$$

where Pr_t is the turbulent Prandtl number, which is selected according to recommendations in [129] or assumed to be equal to one [75].

To solve the system of equations for a turbulent boundary layer, which describes the combined discharge in laser-arc plasma torches operating in the turbulent mode of the plasma gas flow, it is necessary to assign the boundary and initial (input) conditions corresponding to the design of a particular plasma torch. This system of equations consists of four parabolic second- order equations (3.28), (3.29) and (3.10) with regard to variables T, $\overline{w}$, $\overline{u}$ and $A_{1\omega}$, the first-order equation (3.30) to find $\widetilde{v}$ and the elliptical second-order equation (3.33) with regard to H_φ, as well as integral relationships (3.31) and (3.32) which determine the averaged pressure distribution.

The boundary conditions for $\overline{T}$, $\tilde{v}$, $\overline{w}$, $\overline{u}$, H_φ and $A_{1\omega}$ at the axis of the system are normally selected to have the form of (3.13), while those at the external boundary of the calculated domain usually have the form of (3.14) and (3.15), except for the $\tilde{v} = 0$ condition which turns out to be unnecessary within the frames of the approximation used. In addition, condition (3.16) can be used to solve equation (3.10) in the case of a laser beam with an azimuthal variation in the radiation intensity. Initial conditions at the inlet section of the plasma-shaping channel, i.e., at $z = 0$, are assigned in the form of dependencies (3.17). As far as conditions at $z = L$ are concerned, as equations (3.28) to (3.30) and (3.10) are the marching ones on coordinate z, the only condition which should be assigned here is the boundary condition for H_φ. For example, condition of the type of (3.25) can be selected as the condition which describes the laser-arc discharge realised in direct action integrated plasma torches (see Figure 3.1a). In the case of indirect action plasma torches, the calculation domain for equation (3.33) can be limited by the $z = Z_{2a}$ plane (see Item 3.2.1), assuming that $H_\varphi = 0$ at $z \geq Z_{2a}$.

3.2.3 Non-equilibrium model of the combined discharge

The most important restriction of the above-described models of the laser-arc discharge formed using integrated plasma torches is the assumption of the presence of local thermal equilibrium in the plasma. It is a known fact (see, e.g., [75]) that at low pressures and low currents of the arc, near electrodes and cooled walls of the plasma torch channels, the state of the conventional arc plasma can differ greatly from equilibrium. An additional factor that affects the non-equilibrium state of the plasma in the case of the combined discharge is its heating by the laser beam. The mechanism of such heating consists in absorption of the energy of the laser radiation by a lighter electron component of the plasma and then transfer of this energy to heavy particles as a result of collisions. As only a small part of this energy, proportional to the ratio of masses of the colliding particles, is transferred by collisions of electrons with atoms and ions, the laser-arc plasma generated in the devices under consideration can be substantially non-isothermal within its laser beam affected zone, i.e., within the near-axis zone of the discharge (see Figure 3.1).

To describe the non-equilibrium plasma of the combined discharge, let us use the most general model of the ionisation non-equilibrium and non-isothermal plasma. When an atomic gas, e.g., Ar or He, is used as a plasma-forming medium, the plasma can be regarded as a four-component mixture ($\alpha = \overline{1, 4}$) of electrons ($\alpha = 1$), atoms ($\alpha = 2$) and singly ($\alpha = 3$) and doubly charged ($\alpha = 4$) ions. The transport equations for this multicomponent system can be derived within the frames of the 13-momentum approximation of the Grad method [91], where distribution functions of the plasma particles have the following form*:

$$f_\alpha (\mathbf{v}) = n_\alpha \left(\frac{\gamma_\alpha}{2\pi} \right)^{\frac{3}{2}} \exp\left[- \frac{\gamma_\alpha v^2}{2} \right] [1 + \gamma_\alpha w_{\alpha l} v_l + \frac{\gamma_\alpha}{2 p_\alpha} \pi_{\alpha\,ls} (v_l v_s - \frac{1}{3} \delta_{ls} v^2) +$$

$$+ \frac{\gamma_\alpha}{5 p_\alpha} h_{\alpha l} v_l (\gamma_\alpha v^2 - 5)], \quad (\alpha = \overline{1, 4}) . \tag{3.40}$$

Here: $\mathbf{v}$ is the vector of velocity of a particle; $\gamma_\alpha = m_\alpha / (k_B T_\alpha)$; m_α is the mass of a particle of the α kind ($m_1 \equiv m_e$, $m_4 \approx m_3 \approx m_2 \equiv M$ is the mass of an atom of the plasma gas);

---

* The form of writing (3.40) implies summation over repeating Roman subscripts *l* and *s*.

n_α, T_α, $\mathbf{w}_\alpha$, $\hat{\pi}_\alpha$, $\mathbf{h}_\alpha$ are the local values of the concentration, temperature, vector of the diffusion velocity, tensor of viscous stresses and energy flow of particles of the α kind (the first thirteen momenta of the distribution function having an explicit physical meaning); $p_\alpha = n_\alpha k_B T_\alpha$ is the partial pressure of the α-component of the plasma.

In the case of the stationary laser-arc discharge in integrated plasma torches operating in laminar flow mode with a whirling of the plasma gas, the equations of motion and the continuity equation for the entire plasma preserve the form of (3.2) and (3.3), where:

$$\rho = \sum_{\alpha = 1}^{4} m_\alpha n_\alpha$$

is the mass density of the plasma; v, w and u are the radial, azimuthal and axial components of the mean-mass velocity

$$\mathbf{u} = \frac{1}{\rho} \sum_{\alpha = 1}^{4} m_\alpha n_\alpha \mathbf{u}_\alpha$$

($\mathbf{u}_\alpha = \mathbf{u} + \mathbf{w}_\alpha$ is the vector of velocity for the α-component of the plasma);

$$p = \sum_{\alpha = 1}^{4} p_\alpha$$

is the total pressure; j_r and j_z are the radial and axial components of the electric current density

$$\mathbf{j} = \mathbf{j}_e + \mathbf{j}_i$$

($\mathbf{j}_e = - e n_e \mathbf{w}_e$ is the electron current, $\mathbf{j}_i = e \sum_{\alpha = 3,4} Z_\alpha n_\alpha \mathbf{w}_\alpha$ is the ion current in the plasma and Z_α is the charge number for particles of the α kind); $\eta = \sum_{\alpha = 2}^{4} \eta_\alpha$ is the total coefficient of viscosity of the plasma (η_α is the partial viscosity of the α-component) calculated by ignoring the electron viscosity whose contribution to the total viscosity of the mixture under consideration is negligibly small [91].

Assuming that plasma is a two-temperature one, i.e., that $T_e \neq T$, where T_e is the temperature of electrons and $T = T_\alpha$ ($\alpha = 2, 4$) is the temperature of heavy particles, instead of one equation of energy (3.1) we will obtain two equations with regard to T_e and T. Ignoring the energy of the pressure forces and the viscous dissipation of the energy for all components of the plasma, as we did it earlier, the said equations in the chosen cylindrical system of coordinates can be written as follows:

$$\frac{1}{r} \frac{\partial}{\partial r} \left\{ r \left[v \left(\frac{5}{2} n_e k_B T_e + \varepsilon_i \right) + q_{er}^D \right] \right\} + \frac{\partial}{\partial z} \left[u \left(\frac{5}{2} n_e k_B T_e + \varepsilon_i \right) + q_{ez}^D \right] =$$

$$= \frac{1}{r} \frac{\partial}{\partial r} \left(r \chi_e \frac{\partial T_e}{\partial r} \right) + \frac{\partial}{\partial z} \left(\chi_e \frac{\partial T_e}{\partial z} \right) + j_{er} E_r + j_{ez} E_z +$$

$$+ \kappa_\omega \langle S \rangle - \psi - \beta_T (T_e - T); \tag{3.41}$$

$$\frac{1}{r} \frac{\partial}{\partial r} \left[r \left(\rho v \frac{5}{2} \frac{k_B T}{M} + q_r^D \right) \right] + \frac{\partial}{\partial z} \left(\rho u \frac{5}{2} \frac{k_B T}{M} + q_z^D \right) =$$

$$= \frac{1}{r}\frac{\partial}{\partial r}\left(r\chi\frac{\partial T}{\partial r}\right) + \frac{\partial}{\partial z}\left(\chi\frac{\partial T}{\partial z}\right) + j_{ir}E_r + j_{iz}E_z + \beta_T(T_e - T).$$ (3.42)

Here: $\varepsilon_i = n_3(U_2 - \Delta U_2) + n_4(U_2 - \Delta U_2 + U_3 - \Delta U_3)$ is the ionisation energy (U_2 and U_3 are the potentials of ionisation of atoms and singly charged ions of the plasma gas, ΔU_α are the reductions in the corresponding potentials (see Section 2.2) caused by interaction of the charged particles in the plasma); q_e^D, $q^D = \sum\limits_{\alpha=2}^{4} q_\alpha^D$ are the diffusion flows of the energy of electrons and heavy particles (q_α^D is the corresponding partial flow for the α-component); χ_e, $\chi = \sum\limits_{\alpha=2}^{4} \chi_\alpha$ are the coefficients of electron thermal conductivity and thermal conductivity of heavy particles (χ_α is the thermal conductivity of the α-component); β_T is the energy transfer coefficient and the rest of the designations corresponding to those accepted earlier. At elastic collisions of electrons with heavy particles, the energy transfer coefficient can be calculated from the following formula:

$$\beta_T = 3k_B n_e \frac{m_e}{M} \sum\limits_{\alpha=2}^{4} \tau_{e\alpha}^{-1},$$

where:

$$\tau_{\alpha\beta}^{-1} = \frac{16}{3} n_\beta \Omega_{\alpha\beta}^{(1,1)}$$

are the frequencies of elastic collisions of particles of the α and β kinds; $\Omega_{\alpha\beta}^{(1,s)}$ are the Chapman–Cowling integrals determined from expression (2.53), where, for the case of non-isothermal plasma, the factor before the integral and dimensionless velocity of the colliding particles should be replaced, respectively, by

$$\sqrt{\frac{2\pi(\gamma_\alpha + \gamma_\beta)}{\gamma_\alpha \gamma_\beta}} \; ; \; \sqrt{\frac{\gamma_\alpha \gamma_\beta(v_\alpha + v_\beta)^2}{2(\gamma_\alpha + \gamma_\beta)}} \; .$$

The system of equations (3.2), (3.3), (3.41) and (3.42) should be supplemented with equations (3.10) and (3.11) which, along with relationship (3.12), serve for the calculation of a spatial distribution of the intensity of the laser radiation interacting with the electron component of the plasma, as well as with Maxwell's equations (3.4) and (3.5) to find electric and magnetic fields in the discharge. It should be noted here that, as in the case of the non-equilibrium plasma, instead of two Maxwell's equations it is possible to use one equation of the second order for H_φ (see, e.g., Item 3.2.1), which is obtained by excluding E_r and E_z from the said equations using the generalised Ohm's law which relates the current density to the intensity of the electric field.

Composition of the ionisation non-equilibrium plasma can be found as a result of solving continuity equations for individual components:

$$\frac{1}{r}\frac{\partial}{\partial r}(rn_\alpha v_\alpha) + \frac{\partial}{\partial z}(n_\alpha u_\alpha) = \dot{n}_\alpha, \quad (\alpha = \overline{1,4}),$$ (3.43)

where: v_α, u_α are the radial and axial constituents of the velocity of the α-component of the plasma and n_α is the rate of the formation of particles of the α kind per unit volume. Supposing that a change in the number of particles occurs only as a result of ionisation by electron-heavy particle collisions and three-body recombination (ignoring the photoionisation processes), the rates of formation of particles can be written in the following forms:

$$\dot{n}_e = \dot{n}_3 + 2\dot{n}_4; \quad \dot{n}_2 = -K_3 n_e (C_2 n_2 - n_e n_3);$$

$$\dot{n}_3 = -(\dot{n}_2 + \dot{n}_4); \quad \dot{n}_4 = K_4 n_e (C_3 n_3 - n_e n_4), \tag{3.44}$$

where: K_3 and K_4 are the three-body recombination coefficients for singly and doubly charged ions of the plasma gas [130]; C_2 and C_3 are equilibrium constants for the ionisation reactions (right parts of the corresponding equations (2.39)) calculated at the electron temperature. It should be noted that instead of four continuity equations (3.43) it is sufficient to use only two equations by adding to them a relationship for the determination of the total pressure and a condition of quasi-neutrality of the plasma similar to (2.42).

Equations (3.2), (3.3), (3.41) and (3.43) form the basis for the description of the non-equilibrium laser-arc plasma formed by integrated plasma torches, including the plasma of the near-electrode regions of the combined discharge with non-evaporating electrodes. As far as the laser-arc discharge plasma column is concerned, the above system of equations can be somewhat simplified. For example in the energy equation (3.42) it is possible to ignore the diffusion heat flow q^D, as compared with the thermal conductivity of heavy particles [75]. In addition, for the approximate calculation of the composition of the plasma generated in the devices under consideration, it can be regarded as an equilibrium one with regard to the process of second-order ionisation, as the role of the doubly charged ions is most significant in the high-temperature near-axis zone of the discharge, where non-equilibrium of the laser-arc plasma shows itself mainly in its being non-isothermal. Therefore, assuming in (3.43) and (3.44) that $\dot{n}_4 = 0$, to determine the composition of the plasma it is possible to use the continuity equation only for the electron gas, while the concentration of the doubly charged ions can be calculated using the corresponding Saha equation (2.39).

Providing that macroscopic parameters of individual components and of the entire plasma as a whole change little at a distance of about a mean length of the free path of particles, the diffusion velocities and energy flow of electrons, as well as the coefficients of viscosity and thermal conductivity for all components of the plasma, which are included into the simplified system of equations, can be found from the linearised system of equations for the transport coefficients [91]. Assuming that the plasma is isotropic (non-magnetised by its own magnetic field) and separating the electron equations from the equations for heavy particles, we determine the electron transport properties:

$$n_e \mathbf{w}_e = -D_e \nabla n_e - D_{Te} \frac{1}{T_e} \nabla T_e - n_e b_e \mathbf{E} +$$

$$+ n_e \tau_{e0} \sum_{\beta=2}^{4} \left[1 - \left(\frac{6}{5} C_{e\beta} - 1 \right) \alpha_{Te} \right] \tau_{e\beta}^{-1} \mathbf{w}_\beta; \tag{3.45}$$

$$q_e^D = \left[\left(\frac{5}{2} - \alpha_{Te} \right) p_e + \varepsilon_i \right] \mathbf{w}_e + \frac{5}{2} p_e \tau_e^* \sum_{\beta=2}^{4} \left(\frac{6}{5} C_{e\beta} - 1 \right) \tau_{e\beta}^{-1} \mathbf{w}_\beta; \tag{3.46}$$

$$\chi_e = \frac{5}{2} \frac{k_B}{m_e} p_e \tau_e^*, \tag{3.47}$$

where: $D_e = (k_B / m_e) T_e \tau_{e0}$; $D_{Te} = n_e (1 - \alpha_{Te}) D_e$ are the diffusion and thermal diffusion coefficients for electron gas; $b_e = (e / m_e) \tau_{e0}$ is the electron mobility;

$$\tau_{e0}^{-1} = \sum_{\beta = 2}^{4} \left[1 - \left(\frac{6}{5} C_{e\beta} - 1 \right) \alpha_{Te} \right] \tau_{e\beta}^{-1} ;$$

$$\alpha_{Te} = \frac{5}{2} \tau_e^* \sum_{\beta = 2}^{4} \left(\frac{6}{5} C_{e\beta} - 1 \right) \tau_{e\beta}^{-1} \quad \text{is the electron thermal diffusion constant;}$$

$$\tau_e^{*-1} = \frac{2}{5} A_{ee} \tau_{ee}^{-1} + \sum_{\beta = 2}^{4} \left(\frac{5}{2} - \frac{6}{5} B_{e\beta} \right) \tau_{e\beta}^{-1}$$

and the values of $A_{\alpha\beta}$, $B_{\alpha\beta}$ and $C_{\alpha\beta}$ are determined in (2.52). Solving the rest of the system of equations for the transport coefficients of heavy particles ($\alpha = \overline{2, 4}$), we find:

$$n_\alpha w_\alpha = - \sum_{\beta = 2}^{4} D_{\alpha\beta} \nabla n_\beta - D_{T\alpha} \frac{1}{T} \nabla T + n_\alpha b_\alpha \mathbf{E} ; \tag{3.48}$$

$$\eta_\alpha = \frac{1}{2} \sum_{\beta = 2}^{4} \frac{|a|_{\beta\alpha}}{|a|} p_\beta \tau_\beta ; \tag{3.49}$$

$$\chi_\alpha = \frac{5}{2} \frac{k_B}{M} \sum_{\beta = 2}^{4} \frac{|b|_{\beta\alpha}}{|b|} p_\beta \tau_\beta^* , \tag{3.50}$$

where:

$$D_{\alpha\beta} = \frac{|c|_{\beta\alpha}}{|c|} \frac{n_\alpha}{n_\beta} \frac{k_B}{M} T \tau_{\beta 0}$$

are the mutual diffusion coefficients for heavy particles of the α and β kinds;

$$D_{T\alpha} = \sum_{\beta = 2}^{4} n_\beta D_{\alpha\beta} ; \quad b_\alpha = \sum_{\beta = 3}^{4} \frac{|c|_{\beta\alpha}}{|c|} \frac{Z_\beta e}{M} \tau_{\beta 0} ;$$

are the thermal diffusion coefficients and the mobilities for the α kind particles; $|a|$, $|b|$ and $|c|$ are the determinants made up of coefficients

$$a_{\alpha\alpha} = 1; \quad a_{\alpha\beta} = \frac{3}{20} \frac{n_\alpha}{n_\beta} \left(A_{\alpha\beta} - \frac{5}{3} \right) \tau_{\alpha\beta}^{-1} \tau_\alpha, \quad \alpha \neq \beta ,$$

$$b_{\alpha\alpha} = 1; \quad b_{\alpha\beta} = \frac{1}{5} \frac{n_\alpha}{n_\beta} \left(A_{\alpha\beta} - \frac{55}{16} + \frac{3}{4} B_{\alpha\beta} \right) \tau_{\alpha\beta}^{-1} \tau_\alpha^* , \quad \alpha \neq \beta$$

and

$$c_{\alpha\alpha} = 1 ; \quad c_{\alpha\beta} = - \frac{1}{2} \tau_{\alpha\beta}^{-1} \tau_{\alpha 0}, \quad \alpha \neq \beta ;$$

$|a|_{\beta\alpha}$ and $|b|_{\beta\alpha}$ are the algebraic adjuncts of elements $\beta\alpha$;

$$\tau_{\alpha 0}^{-1} = \frac{1}{2} \sum_{\substack{\beta = 2 \\ \beta \neq \alpha}}^{4} \tau_{\alpha\beta}^{-1} ;$$

$$\tau_{\alpha}^{-1} = \frac{3}{10} A_{\alpha\alpha} \tau_{\alpha\alpha}^{-1} + \frac{3}{20} \sum_{\substack{\beta = 2 \\ \beta \neq \alpha}}^{4} \left(A_{\alpha\beta} + \frac{5}{3} \right) \tau_{\alpha\beta}^{-1} ;$$

$$\tau_{\alpha}^{*-1} = \frac{2}{5} A_{\alpha\alpha} \tau_{\alpha\alpha}^{-1} + \frac{1}{5} \sum_{\substack{\beta = 2 \\ \beta \neq \alpha}}^{4} \left(A_{\alpha\beta} + \frac{55}{16} - \frac{3}{4} B_{\alpha\beta} \right) \tau_{\alpha\beta}^{-1} .$$

Total coefficients of viscosity of the plasma and thermal conductivity of a heavy component are derived by summing up expressions (3.49) and (3.50) with respect to $\alpha = \overline{2, 4}$.

To determine the values of κ_{ω} and ψ which are included in the electron energy equation, it is possible to use relationships (2.57) and (2.58) by assuming that $T = T_e$ and by summing up with respect to $\alpha = 2, 3$. Finally, as far as the real and imaginary parts of the complex dielectric permittivity of the plasma included in equations (3.10) and (3.11) are concerned, expressions (2.55) and (2.56) remain valid for them.

The above-described system of equations for modelling the non-equilibrium plasma of the laser-arc discharge generated in integrated plasma torches with laminar flow of the atomic plasma gas, differs from the LTE model of the combined discharge, considered in Item 3.2.1, in the availability of two additional equations (equations of energy (3.41) and continuity (3.43) for electron gas). Therefore, consider assignment of the boundary conditions only for the said equations, as the corresponding conditions for temperature T of the heavy component of the plasma, gas-dynamic variables v, w, u, p and complex amplitudes of the laser radiation field $A_{(1, 2)\,\omega}$ remain as they are for the equilibrium model (see Item 3.2 a).

Before determining the boundary conditions required for solving equations (3.41) and (3.43), transform the continuity equation for the electron gas to reduce it to a more convenient form. Using the determination of the total velocity of the electron component of the plasma and the expression for the diffusion flow of electrons (3.45), yields, instead of (3.43), a differential equation of the second order with respect to n_e. Therefore, we have to assign boundary conditions for two second order differential equations.

The choice of the corresponding conditions at the discharge axis is based on the assumption of cylindrical symmetry of distributions of T_e and n_e:

$$\frac{\partial T_e}{\partial r} = 0; \quad \frac{\partial n_e}{\partial r} = 0 . \tag{3.51}$$

It is assumed that at the external boundary of the plasma (both free and in contact with the plasma torch channel wall), i.e. at $r = R_{\sigma}$, where $R_{\sigma}(z)$ is the radius of the current-conducting region, the electric current passing through this boundary is equal to zero and, hence, diffusion of the charged particles occurs by an ambipolar mechanism. Assuming also that near the external boundary of the discharge the plasma is slightly ionised and contains only singly charged ions, the boundary conditions for n_e and T_e at $r = R_{\sigma}$ can be found from the following system of equations:

$$n_e w_{en}^A = \frac{1}{4} v_{Te} \exp\left(\frac{e \Delta \varphi}{k_B T_e} \right) ;$$

$$- \lambda_e \nabla_n T_e + \left\{ \frac{5}{2} \left[1 - \tau_e^* \left(\frac{6}{5} C_{e2} - 1 \right) \tau_{e2}^{-1} \right] - \alpha_{Te} \right\} p_e w_{en}^A =$$

$$= \frac{1}{4} v_{Te} \exp \left(\frac{e \Delta \varphi}{k_B T_e} \right) p_e \left(2 - \frac{e \Delta \varphi}{k_B T_e} \right).$$

(3.52)

Here: w_{en}^A is the component of the ambipolar diffusion velocity of electron normal to the plasma boundary;

$$v_{Te} = \sqrt{8 k_B T_e / \pi m_e}$$

is the electron thermal speed;

$$\Delta \varphi = \frac{k_B T_e}{e} \ln \sqrt{T m_e / M T_e}$$

is the jump of the potential ($\Delta \varphi < 0$) occurring in the collision-free layer of the space discharge adjoining the boundary of the current-conducting region of the plasma, the position of this boundary being determined by condition (3.8). To find the ambipolar diffusion velocity of electrons, in the case under consideration, it is possible to use the following expression [91]:

$$n_e w_e^A = - D_e^A \nabla n_e - D_{Te}^A \frac{1}{T_e} \nabla T_e - D_i \frac{1}{T} \nabla T,$$

(3.53)

where:

$$D_e^A = D_i \left(1 + \frac{T_e}{T} \right)$$

is the coefficient of ambipolar diffusion of electrons;

$$D_{Te}^A = n_e (1 - \alpha_{Te}) D_i \frac{T_e}{T}$$

is the coefficient of ambipolar thermal diffusion of electrons;

$$D_i = \frac{2 k_B T}{M} \tau_{32}$$

is the coefficient of diffusion of ions in the slightly ionised plasma.

To describe the discharge in indirect action integrated plasma torches, where the plasma-shaping nozzle also serves as the anode of the arc (see Figure 3.1b–d), within the zone of the anode fixation of the discharge, i.e. at $r = R_C$; $Z_{1a} < z < Z_{2a}$ (see Item 3.2.1), instead of conditions (3.52) it is necessary to use, for example, the following conditions:

$$T_e = T_{ea}; \quad n_e = n_{ea},$$

(3.54)

where: T_{ea} and n_{ea} are the temperature and concentration of electrons within the anode region, which are assigned on the basis of the available experimental data and results of theoretical calculations [119, 120]. It should be noted here that conditions (3.52) can still be used for the external boundary of the currentless region of the plasma ($z > Z_{2a}$) generated in the devices under consideration.

Boundary conditions at the inlet section of the calculated region, i.e. at $z = 0$, for equations (3.41) and (3.43) are normally assigned in the form of the following distributions:

$$T_e = T_e(r, 0) \; ; \; n_e = n_e(r, 0). \tag{3.55}$$

As the inlet section of the channel is located as a rule near the cathode, the explicit form of these distributions should be determined with allowance for peculiarities of the cathode processes for a particular design of the cathode unit of the laser-arc plasma torch.

As far as the boundary conditions in the outlet plane of the calculated region, i.e. at $z = L$, are concerned, because this plane is usually selected to be in the immediate vicinity of the anode (workpiece) surface, for consideration of the laser-arc discharge in the direct action integrated plasma torches (see Figure 3.1a) conditions (3.54) should be used in this case for n_e and T_e. Finally, if we consider the plasma jet which is generated in indirect action laser-arc plasma torches (see Figure 3.1b–d) and which flows out into a free space, or if it is assumed that the surface of a workpiece is located sufficiently far away from the outlet section of the plasma torch (outside the calculation domain), instead of conditions (3.54) at $z = L$ it is possible to use, for example, the "soft" boundary conditions (3.23), where $\phi = \{n_e, T_e\}$.

3.2.4 Model of the cathode phenomena for the thermionic tubular cathode of integrated plasma torch

As already noted, the correct statement of boundary conditions near the cathode for all of the above models of the combined discharge generated by various laser-arc plasma torches is impossible without a detailed investigation into the cathode phenomena as applied to a particular design of cathode unit of the integrated plasma torch. For example, consider a plasma torch with a tubular refractory cathode (see Figure 3.1b), the design of which is shown in detail in Figure 3.2. To develop a model of the cathode processes for such a cathode operating in the thermionic mode, we will use the most general approach consisting of a self-consistent allowance for the entire set of interrelated physical phenomena occurring in the body of the cathode, on its surface and in the near-cathode plasma [119, 131].

According to [132], divide conditionally the cathode region of the discharge into two layers: the quasi-neutral ionisation layer, in which generation of the charged particles takes place, and the space charge layer on which the major part of the cathode potential drop falls. The space charge layer adjoining the cathode surface has a thickness of the order of the Debye radius and can be considered to be collisionless, as, under a pressure close to atmospheric, the free path of particles is much larger than r_D [120]. Electrons emitted from the cathode are accelerated by the electric field of the space charge and acquire energy sufficient for ionisation of atoms of the plasma gas in a collision ionisation layer. The formed ions and high-energy electrons of the plasma, which are capable of overcoming the cathode potential drop (the so-called "reversed" electrons) come to the surface of the cathode and transfer to it the energy required for heating the cathode and generating the thermoelectronic emission current.

In addition to the energy fed through the arc spot, the tubular cathode design under consideration allows part of the power of the laser beam transmitted via the hole in the cathode to be used for additional heating of the internal surface of its tip (see Figure 3.2). For this purpose, geometrical parameters of this surface should be selected so that peripheral rays of the beam, being reflected from the internal conical surface, get into the cylindrical outlet and have a sufficient number of reflections to be fully absorbed by the cathode material.

To calculate the resultant temperature field of the cathode, which is assumed to be axially symmetrical, we will use a stationary heat conduction equation with allowance for the Joule heat release:

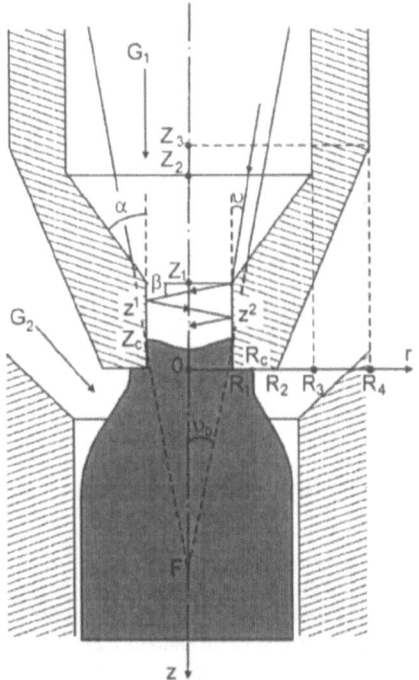

Figure 3.2 Diagram of the refractory tubular cathode for laser-arc plasma torches.

$$\frac{1}{r}\frac{\partial}{\partial r}\left(r\chi_c\frac{\partial T^c}{\partial r}\right) + \frac{\partial}{\partial z}\left(\chi_c\frac{\partial T^c}{\partial z}\right) + \rho_c j_c^2 = 0 .\tag{3.56}$$

Here: $T^c(r, z)$ is the cathode temperature; $\chi_c(T^c)$ and $\rho_c(T^c)$ are the coefficients of thermal conductivity and specific electrical resistance of the cathode material; $j_c(r, z) = -\nabla\Phi^c/\rho_c$ is the current density in the cathode body where distribution of the potential $\Phi^c(r, z)$ can be found from the equation of continuity of the current in metal:

$$\frac{1}{r}\frac{\partial}{\partial r}\left(\frac{r}{\rho_c}\frac{\partial\Phi^c}{\partial r}\right) + \frac{\partial}{\partial z}\left(\frac{1}{\rho_c}\frac{\partial\Phi^c}{\partial z}\right) = 0 .\tag{3.57}$$

Boundary conditions for equations (3.56) and (3.57) on the cathode surface are set based on the following considerations. Assume that, the refractory cathode being investigated is pressed into a copper water-cooled yoke, then at $z = -L_w$; $R_3 \le r \le R_4$ (L_w is the cathode length) it can be assumed that:

$$T^c = T_w; \quad \Phi^c = 0,\tag{3.58}$$

where T_w is the cooling water temperature. Conditions of convective-radiative heat exchange with an ambient gas and the absence of the current density component normal to the cathode surface are assumed to hold on the external surface of the cathode, outside the arc fixation zone. Therefore, at $r = R_4$; $-L_w \le z < Z_3$, at $r = R_2 + [(R_4 - R_2) z/Z_3]$; $Z_3 \le z < 0$ and at $R_c < r \le R_2$; $z = 0$ (see Figure 3.2), the boundary conditions have the following form:

$$- \chi_c \nabla_n T^c = \alpha_{c2} (T^c - T_2) + \varepsilon_c \sigma_0 (T^{c^4} - T_2^4); \; \nabla_n \Phi^c = 0, \qquad (3.59)$$

where: α_{c2} is the heat-transfer coefficient; T_2 is the external plasma gas temperature; ε_c is the degree of blackness of the cathode material and σ_0 is the Stefan-Boltzmann constant. The boundary conditions on the internal cathode surface outside the zone of contact with the plasma, i.e. at $r = R_3; -L_w \le z < Z_2$, at $r = R_1 + [(R_3 - R_1)(z - Z_1)/(Z_2 - Z_1)]$; $Z_2 \le z < Z_1$ and at $r = R_1; Z_1 \le z < Z_c$ (see Figure 3.2), are selected similarly to those in (3.59):

$$- \chi_c \nabla_n T^c = - \alpha_{c1} (T^c - T_1) - \varepsilon_c \sigma_0 (T^{c^4} - T_1^4); \; \nabla_n \Phi^c = 0, \qquad (3.60)$$

where: α_{c1} is the corresponding heat-transfer coefficient and T_1 is the internal plasma gas temperature*.

Boundary conditions within the zone of the cathode fixation of the discharge (see Figure 3.2) are set as follows:

$$- \chi_c \frac{\partial T^c}{\partial z} = - q_c^a; \; - \frac{1}{\rho_c} \frac{\partial \Phi^c}{\partial z} = - j_c, \; \text{at} \, R_1 \le r \le R_c; z = 0 \qquad (3.61)$$

and

$$- \chi_c \frac{\partial T^c}{\partial r} = q_c^a; \; - \frac{1}{\rho_c} \frac{\partial \Phi^c}{\partial r} = j_c, \; \text{at} \, r = R_1; Z_c \le z \le 0, \qquad (3.62)$$

where: q_c^a and j_c are the heat flow introduced to the cathode by the arc and the current density at the cathode which, following [133, 134], are assumed to be distributed over the cathode spot region. To take into account the additional laser heating of the internal cathode surface, boundary conditions for the temperature in (3.60) and (3.62) should be corrected by adding to the right parts of the said conditions the q_c^l term, describing the distribution of the laser radiation power absorbed by the internal conical surface of the cathode and its outlet surface.

Distribution of the arc heat flow $q_c^a (r_s, z_s)$ over the cathode spot surface (r_s and z_s are the coordinates of the point on this surface) can be found from the known relationship [131, 135]:

$$q_c^a = j_i \left(V_c + U_2 - \overline{\varphi}_c + \frac{2k_B T}{e} \right) + j_e \left(\overline{\varphi}_c + \frac{2k_B T_e}{e} \right) - j_{em} \left(\overline{\varphi}_c + \frac{2k_B T^c}{e} \right). \qquad (3.63)$$

Here: $j_i (r_s, z_s)$ and $j_e (r_s, z_s)$ are the local values of density of the ion and electron currents flowing from the plasma to the cathode; $j_{em} (r_s, z_s)$ is the distribution of the current density of the emitted electrons over the spot surface; $V_c (r_s, z_s)$ is the local value of the cathode potential drop; U_2 is the potential of ionisation of the plasma gas atoms (the near-cathode plasma is assumed to contain only the singly charged ions); $T_e (r_s, z_s)$ and $T (r_s, z_s)$ are the distributions of temperatures of electrons and heavy particles of the plasma along the boundary of an ionisation layer with the space charge layer; $T^c (r_s, z_s)$ is the local value of the cathode surface temperature and $\overline{\varphi}_c (r_s, z_s)$ is the effective work function of the electrons with Schottky's correction:

$$\overline{\varphi}_c = \varphi_c - \sqrt{eE_c/4\pi\varepsilon^0}, \qquad (3.64)$$

where: φ_c is the work function of the cathode material; $E_c (r_s, z_s)$ is the electric field strength at the cathode surface which in the case of the collisionless layer of the space charge is

---

* The values of α_{c1} and α_{c2}, as well as the corresponding values of T_1 and T_2 in (3.59) and (3.60), should be calculated depending upon the flow rates of G_1 and G_2, the methods of feeding and the thermal-physical properties of the internal and external plasma gases.

determined either by using MacKeown's equation [136] or by solving a more precise equation which allows for "hot" electrons present in this layer, coming from the plasma. As under a pressure close to atmospheric, the thickness of the space charge layer is substantially smaller than the linear dimensions of the region of the cathode fixation of the discharge and its curvature radius R_1 (see Figure 3.2), this layer can be considered to be all flat, and the one-dimensional Poisson's equation can be used to find E_c (r_s, z_s) and V_c (r_s, z_s) [135]:

$$\varepsilon^0 \frac{d^2\varphi}{dx^2} = \frac{j_{em}}{\sqrt{\frac{2e}{m_e}\varphi + \frac{8k_BT^c}{\pi m_e}}} - \frac{j_i}{\sqrt{\frac{2e}{M}(V_c - \varphi) + \frac{8k_BT}{\pi M}}} +$$

$$+ j_e \sqrt{\frac{\pi m_e}{2k_BT_e}} \left[1 + \frac{2}{\sqrt{\pi}} \int_0^{\sqrt{\frac{e(V_c - \varphi)}{k_BT_e}}} e^{-t^2} dt \right] \exp\left[-\frac{e(V_c - \varphi)}{k_BT_e} \right], \tag{3.65}$$

with boundary conditions:

$$\varphi\,(r_s, z_s) = 0, \text{ at } x = 0 \tag{3.66}$$

and

$$\varphi\,(r_s, z_s) = V_c - \frac{k_BT_e}{2e}; \quad \frac{d\varphi}{dx} = \frac{3k_BT_e}{32e}\sqrt{(\pi M/k_BT)}\,\tau_{32}^{-1}, \text{ at } x = l_1, \tag{3.67}$$

where: $\varphi\,(x, r_s, z_s)$ is the spatial distribution of the potential in the space charge layer; x is the distance from the cathode surface; l_1 is the thickness of the space charge layer, which is determined using the second condition in (3.67) and τ_{32}^{-1} is the frequency of ion-atom collisions determined in Item 3.2.3.

Distribution of the current density j_c (r_s, z_s) in the cathode region is calculated using the following relationship:

$$j_c = j_i - j_e + j_{em} . \tag{3.68}$$

Here:

$$j_i\,(r_s, z_s) = en_e \sqrt{k_BT_e/M} \, \exp\left[-\frac{1}{2} \right]$$

is the local current density of the ions from the near-cathode plasma [137];

$$j_e\,(r_s, z_s) = \frac{1}{4}\alpha_e en_e \sqrt{8k_BT_e/\pi m_e} \, \exp\left[-\frac{eV_c}{k_BT_e} \right]$$

is the local current density of the "reversed" electrons [135];

$$j_{em}\,(r_s, z_s) = A_R T^{c^2} \exp\left[-\frac{e\overline{\varphi}_c}{k_BT^c} \right]$$

is the distribution of the electron emission current on the cathode spot surface; α_e is the coefficient of accommodation of the electrons [138]; n_e (r_s, z_s) is the electron density distribution on the boundary of the quasi-neutral ionisation layer and the space charge layer; A_R is Richardson's constant. In addition, distribution of the total current density at the cathode (3.68) on its surface should meet the explicit integral relationship:

$$I = 2\pi \left[\int_{R_1}^{R_c} j_c (r_s, 0) \, r_s \, dr_s + R_1 \int_0^{Z_c} j_c (R_1, z_s) \, dz_s \right].$$

(3.69)

To find distribution of the concentration of electrons of the plasma along the boundary of the ionisation layer and the space charge layer, an assumption of the partial local thermal equilibrium present here can be used as the first approximation [134]. In this case, the composition of the three-component plasma under consideration can be determined using the Saha equation (2.39) ($\alpha = 2$), taken at an electron temperature, the equation of state:

$$p = n_e k_B T_e + (n_2 + n_3) \, k_B T$$

(3.70)

and the quasi-neutrality condition for the plasma:

$$n_e = n_3 \, ,$$

(3.71)

where: $p \, (r_s, z_s)$ is the pressure of the plasma gas near the cathode surface, the rest of the designations corresponding to those assumed earlier.

To have a more accurate determination of distributions of the concentrations of particles within the ionisation layer, generally speaking, allowance should be made for ionisation non-equilibrium of the plasma [131, 135]. In this case, instead of the Saha equation, it is necessary to solve the continuity equation (3.43) for one, e.g. ion, component of the plasma and use relationships (3.70) and (3.71). Assuming that variation in the concentration of particles across the ionisation layer is more substantial than along the boundary of the layers and that its thickness is much smaller than R_1, the existing one-dimensional continuity equation can be written in a similar form both for internal ($r \le R_1$; $Z_c \le z \le 0$) and external ($R_1 \le r \le R$; $z \ge 0$) regions of the cathode fixation of the discharge (see Figure 3.2):

$$\frac{d}{dx} (n_3 u_{3n}) = \dot{n}_3 \, .$$

(3.72)

Here: $u_{3n} \, (r_s, z_s)$ is the distribution of the component of velocity of ions in the internal and external regions of the ionisation layer normal to the corresponding part of the cathode surface; x is the distance from the boundary of this layer with the space charge layer, and the value of $\dot{n}_3$ determined in (3.44) for the case of single ionisation under consideration is $K_3 n_e \, (C_2 n_2 - n_e n_3)$. The boundary condition for integration of equation (3.72) at $x = 0$ in an assumption that an infinitely thin space charge layer can be assigned directly at the cathode surface:

$$n_3 \, (r_s, z_s) \, u_{3n} \, (r_s, z_s) = - \frac{j_i}{e} \, ,$$

(3.73)

while the thickness of the ionization layer l_2, can be determined, e.g., as the distance from the cathode surface in which the following condition is met [135]:

$$\frac{C_2 n_2 - n_e^2}{C_2 n_2} = 0.05 \, ,$$

(3.74)

i.e. the plasma is assumed to be almost equilibrium in terms of ionisation.

Being part of equation (3.72), the velocity of the ion component of the plasma in the ionisation layer can be found from the system of equations of motion for all components, allowing for the inertia terms [131, 135]:

$$m_\alpha n_\alpha u_{\alpha n} \nabla_n u_{\alpha n} + \nabla_n p_\alpha - e Z_\alpha n_\alpha E_n =$$

$$= - n_\alpha \sum_{\beta=1}^{3} \mu_{\alpha\beta} \tau_{\alpha\beta}^{-1} (u_{\alpha n} - u_{\beta n}), \quad (\alpha = \overline{1, 3}),$$

(3.75)

where the designations assumed in Item 3.2.1 are used. Boundary conditions for equations (3.75) are assigned in the following form:

$$n_2 (r_s, z_s) u_{2n} (r_s, z_s) = \frac{j_i}{e}, \quad \text{at } x = 0;$$

$$u_{en} (r_s, z_s) = - b_e E_n; \quad u_{3n} (r_s, z_s) = b_3 E_n, \quad \text{at } x = l_2,$$

(3.76)

where: b_e and b_3 are the mobilities of electrons and ions (see Item 3.2.3), and the distribution of the electric field component within the ionisation layer, which is normal to the cathode surface, is determined from the following condition:

$$n_e u_{en} - n_3 u_{3n} = \frac{j_c}{e}.$$

(3.77)

The temperatures of electrons and heavy particles in the ionisation layer can be considered to be constant through its thickness with sufficient accuracy. Besides, $T (r_s, z_s)$ is normally assumed to be equal to the local value of the cathode surface temperature and $T_e (r_s, z_s)$ is found from the equation of the energy balance in this layer [135]:

$$j_{em} \left(V_c + \frac{2k_B T^c}{e} \right) = j_i \left(U_2 + \frac{2k_B T}{e} \right) +$$

$$+ j_e \left(V_c + \frac{2k_B T_e}{e} \right) + j_c \left(\frac{5}{2} - \alpha_{Te} \right) \frac{k_B T_e}{e},$$

(3.78)

where: α_{Te} is the electron thermal diffusion constant determined in Item 3.2.3. The term in the left part of equation (3.78) describes the energy brought to the ionisation layer by emission electrons, the first two terms in the right part of this equation describe the energy removed to the cathode by ions and "reversed" electrons of the plasma and, finally, the last term describes the energy removed to the discharge column. As to the energy of the laser radiation (both passed via the hole in the cathode and reflected from its internal surface), which can be put into the plasma of the ionisation layer within the internal region of the cathode fixation of the discharge (see Figure 3.2), it turns out to be low for the cathode design under consideration and because of the small thickness of this layer can be ignored in equation (3.78).

Distributions of $T_e (r_s, z_s)$ and $T (r_s, z_s)$ thus determined serve not only for closure of the described model of the processes occurring in the near-cathode region, but also are the boundary conditions for the non-equilibrium model of the combined discharge plasma column at the boundary with the cathode (see Item 3.2.3). It should be noted here that, to simplify the model of the cathode phenomena under investigation, the temperatures of electrons and heavy particles in the ionisation layer can be considered to be identical [134] and equal to the common temperature of the plasma near the cathode T_c and this temperature can be used as the boundary condition for the LTE-model of the combined discharge plasma column (see Items 3.2.1, 3.2.2).

As was noted above, the suggested design of the thermionic cathode of the laser-arc plasma torches makes it possible to use peripheral rays of the laser beam passed through the cathode for additional heating of its working surface (see Figure 3.2). Prior to determining distribution of the radiation power absorbed by the internal cathode surface q_c^1, we will consider the condition under which all the laser radiation reflected from a conical portion of this surface comes to an outlet of the cathode. If $F^2 >> [\lambda/(\pi \vartheta_b^2)]^2$, the laser beam in the $z < 0$ range can be assumed to be a set of the straight rays spherically converging to the point of intersection of the focal plane with the beam axis, each of these rays being characterised by a peculiar value of the ϑ angle (see Figure 3.2). Determining the angle of focusing of the beam ϑ_b, as the angle within which, for example, 99 % of the laser radiation power is distributed, and considering this angle to be small (tg $\vartheta_b \approx \vartheta_b$), the condition to be found for the rays falling on the internal conical surface of the cathode can be written as follows:

$$R_1 \left[(\text{tg}\alpha - 2\vartheta) \, \text{tg}\beta + 1\right] +$$
$$+ (F - Z_1)(\text{tg}\alpha \, \text{tg}\beta - 1) \, \vartheta \geq 0, \quad \left(\frac{R_1}{F - Z_1} \leq \vartheta \leq \vartheta_b\right)$$

(3.79)

and the positions of the conjugate spots for the reflected rays in the cathode outlet (see Figure 3.2) in this case can be determined using recurrent relationships:

$$z^1(\vartheta) = Z_1 + \frac{R_1\left[(\text{tg}\alpha - 2\vartheta)\,\text{tg}\beta + 1\right] + (F - Z_1)(\text{tg}\alpha\,\text{tg}\beta - 1)\,\vartheta}{\text{tg}\alpha - \vartheta};$$
$$z^{i+1}(\vartheta) = z^1(\vartheta) + 2R_1\,\text{tg}\beta, \quad (i = 1, 2, 3, ...),$$

(3.80)

where:

$$\beta(\vartheta) = \frac{\pi}{2} - 2\alpha + \vartheta$$

is the angle of inclination of the reflected rays relative to the horizontal plane. A condition that these rays undergo not less than n reflections in the outlet channel has the following form:

$$z^n(\vartheta) \leq 0, \quad \left(\frac{R_1}{F - Z_1} \leq \vartheta \leq \vartheta_b\right),$$

(3.81)

and the number of reflections within the channel required for their effective absorption (e.g., not less than 90%) by the cathode surface can be estimated using the following relationship:

$$(1 - \Gamma_\omega)^{n+1} \leq 0.1,$$

(3.82)

where: Γ_ω is the coefficient of absorption of the given wave length of the laser radiation by the cathode material surface*.

To meet condition (3.79), distribution of the heat flow $q_c^1(R_1, z)$ introduced into the body of the cathode by the laser radiation reflected from its internal surface along the length of the outlet of the cathode ($Z_1 \leq z \leq 0$) can be written as follows:

---

* As far as the dependence of this coefficient upon the angle of incidence, polarisation of the incident radiation and temperature of the metal surface is concerned [53, 110], since the β angle is as a rule small, here and further on as Γ_ω we can use the polarisation-independent value of the absorption coefficient at normal incidence, corresponding to a certain temperature of the surface.

$$q_c^j = \int\limits_{\frac{R_1}{F-Z_1}}^{\vartheta_b} \sum_{i=1}^{\infty} \delta\left[z - z^i(\vartheta)\right]\theta\left[-z^i(\vartheta)\right] S^i(\vartheta)\cos\beta\,\Gamma_\omega^i(\vartheta)\,d\vartheta.$$

(3.83)

Here: $\delta(x)$ is the delta function;

$$\theta(x) = \begin{cases} 1, & \text{at } x \geq 0, \\ 0, & \text{at } x < 0; \end{cases}$$

$S^i(\vartheta)$ is the incident laser radiation intensity for the corresponding conjugate spot of the outlet channel surface; $\Gamma_\omega^i(\vartheta) \equiv \Gamma_\omega\{T^c[R_1, z^i(\vartheta)]\}$ are the values of the coefficient of absorption of the laser radiation by the cathode material at the same spot. Ignoring absorption of the reflected rays in the plasma at $z \geq Z_c$, values of the incident laser radiation intensity for all conjugate spots $z^i(\vartheta)$ can be found using the following recurrent relationships:

$$S^1(\vartheta) = S^0(\vartheta)\frac{\text{tg}\alpha}{\cos\beta}R_\omega(\vartheta)\frac{[(F-Z_1)\,\text{tg}\alpha - R_1]^2\,\vartheta}{R_1 z_\vartheta^1 (\text{tg}\alpha - \vartheta)^3};$$

$$S^{i+1}(\vartheta) = S^i(\vartheta)\left[1 - \Gamma_\omega^i(\vartheta)\right]\frac{z_\vartheta^i}{z_\vartheta^{i+1}}, \quad (i = 1, 2, 3, \ldots),$$

(3.84)

where: $S^0(\vartheta)$ is the intensity of the radiation of the initial laser beam $S^0(r, z)$ at the point of incidence of the ray ϑ on the internal conical surface of the cathode

$$r = \frac{[(F-Z_1)\,\text{tg}\alpha - R_1]\,\vartheta}{\text{tg}\alpha - \vartheta}; \quad z = \frac{Z_1\text{tg}\alpha - F\vartheta + R_1}{\text{tg}\alpha - \vartheta};$$

$R_\omega(\vartheta)$ is the coefficient of reflection of the laser radiation at the same point of the said surface; $z_\vartheta^i \equiv dz^i/d\vartheta$. In these designations, for the heat flow of $q_c^j(\vartheta)$ introduced by the laser radiation through the conical portion of the internal cathode surface ($z < Z_1$) at the point of incidence of the ray, ϑ, it holds:

$$q_c^j = S^0(\vartheta)\sin\alpha\left[1 - R_\omega(\vartheta)\right].$$

(3.85)

It should be noted here that distributions in (3.83) and (3.85) are assumed to be axially symmetrical; therefore, for laser beams with azimuthal variations of the radiation field the intensity distribution averaged over the ϑ angle (3.9) should be used to determine the $S^0(\vartheta)$ dependence. As far as the reflection coefficient $R_\omega(\vartheta)$ is concerned, as the angle of incidence of the rays on the internal conical surface, which is approximately equal to $(\pi/2) - \alpha$, is not small, this coefficient will depend not only upon the temperature of the surface at the point of incidence of the ray, but also upon the radiation polarisation in the initial laser beam. To avoid the resulting azimuthal dependence q_c^j, the corresponding value of the reflection coefficient for the non-polarised radiation [110], having $(\pi/2) - \alpha$ incidence angle, can be used as $R_\omega(\vartheta)$ in (3.84) and (3.85).

This covers the description of the closed model of the cathode phenomena for the tubular thermionic cathode of laser-arc plasma torches, allowing for additional heating of the cathode by the laser beam passed through it.

3.3 Simulation of the Integrated Plasma Torch for Laser-Plasma Powder Deposition

As an example of the practical application of integrated plasma torches described in Section 3.1, consider the process of powder deposition with an additive powder fed to the plasma of the combined discharge formed using a specialised laser-arc plasma torch. Mathematical modelling was carried out and a prototype of the integrated plasma torch for laser-plasma surfacing was designed and manufactured to realise this process. The nozzle part of this plasma torch is schematically shown in Figure 3.3.

In the device suggested, the DC arc is burning in an axial flow of the plasma gas (Ar) between a refractory (W) tubular cathode 1 and a workpiece (anode) 2. In the initial region, the discharge is stabilised by the wall of the plasma-shaping nozzle 3, which is coaxial with the cathode. Focusing nozzle 4 provides a distributed feed of the additive powder to the discharge. Both nozzles are made from copper and cooled by water 5. The optically focused beam 6 radiated from the CW CO_2-laser is introduced into the discharge via the hole in the cathode and propagates along the axis of the plasma torch. Plasma gas 7, 8 is fed to the plasma-shaping channel both into the cathode hole and into the gap between the cathode and the wall of nozzle 3 with a possibility of varying the corresponding gas flows independently of each other. Carrier gas 9 is fed into the gap between the plasma-shaping nozzle and nozzle 4. This carrier gas is also argon, and it is used to feed the additive material to the discharge. An open region of the discharge (outside the plasma torch) is blown with a flow of shielding gas (Ar) and the outside environment is under atmospheric pressure.

This laser-arc plasma torch was designed for operation at arc currents of 100 A $\leq I \leq$ 300 A with a laser beam having the TEM_{20} mode, power of $Q_0 \leq$ 5 kW and beam focusing

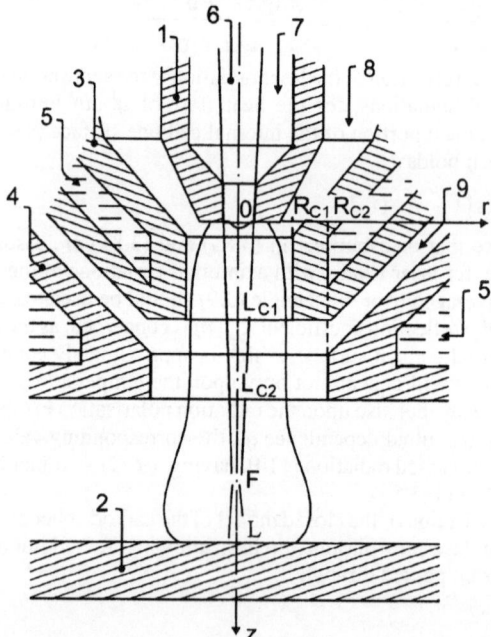

Figure 3.3 Schematic drawing of the nozzle part of the integrated plasma torch for laser- plasma powder deposition: 1, tubular cathode; 2, anode (workpiece); 3, plasma-shaping nozzle; 4, powder focusing nozzle; 5, cooling water; 6, focused laser beam; 7, 8, plasma-forming gas; 9, transport gas.

angle determined, as is indicated in Item 3.2.4, $\vartheta_b = 0.053$. In the development of the cathode unit of the plasma torch, the following external dimensions were selected for the refractory tubular cathode (see Figure 3.2): $R_2 = 2.0$ mm; $R_4 = 4.5$ mm; $Z_3 = -5.0$ mm; $L_w = 20$ mm and provision was made for regulation of heating its tip by laser radiation. For this, the radius of the cylindrical outlet hole of cathode R_1 was assumed to be equal to 1 mm, and the distance from the $z = Z_1$ plane (see Figure 3.2) to the focus of the initial beam $f = F - Z_1$, was taken to vary within the range of 16 mm $\leq f \leq$ 24 mm. For the above values of R_1 and ϑ_b, a marked heating of the cathode with laser radiation will be observed at $f \geq 19$ mm (the dependence of the laser radiation power falling on the internal surface of the cathode ΔQ^0, upon its position with respect to the focal plane for the TEM_{20} beam is shown in Figure 3.4). It should be noted here that there is no point in further increasing the distance between the focus of the initial beam and the cathode ($f > 24$ mm), as this, along with growth of ΔQ^0, will lead to a substantial diffraction distortion of the beam that has passed.

The angle of opening of the internal conical surface of the cathode α, and the length of its outlet hole $-Z_1$ (see Figure 3.2) were chosen on the basis of optimal utilisation of power taken from the laser beam for heating the surface of the outlet channel of the cathode. For this we used conditions (3.79), (3.81) and (3.82), implying that peripheral rays of the laser beam reflected by the conical part of the internal surface of the cathode come to the outlet hole and undergo there a sufficient number of reflections to be effectively absorbed. As the coefficient of absorption of laser radiation with a wave length of $\lambda = 10.6$ μm by the surface of tungsten at a temperature of 2000 K to 3000 K and normal incidence is $\Gamma_\omega \approx 0.1$ [53], there should be not less than 20 reflections of the rays within the outlet channel of the cathode to meet condition (3.82). On this basis, we determined the optimal (for the entire range of variations in f) values of geometrical parameters of the internal surface of the cathode: $\alpha = 42.5°$; $Z_1 = -2.0$ mm; $R_3 = 3.0$ mm.

The detailed numerical investigation into cathode phenomena using the mathematical model described in Item 3.2.4* was conducted for the selected geometry of the tubular cathode of the laser-arc plasma torch under consideration. Activated tungsten ($W + Y_2O_3$ (1 %)) with a work function of $\varphi_c = 3.3$ eV [139] was used as the cathode material. The temperature dependencies of thermal conductivity χ_c, specific electrical resistance ρ_c, blackness degree ε_c and coefficients of absorption Γ_ω (at normal incidence) and reflection R_ω (at an incidence angle of 47.5°) of a non-polarised CO_2-laser radiation for tungsten, used for the calculations, were determined on the basis of data given in [140, 141] and are summarised in Table 3.1.

The flow rate of argon blown through the cathode hole was assumed to be equal to 0.5 l/min at standard condition, i.e. $G_1 = 0.0135$ g/s, while that blown through the gap between the cathode and plasma-shaping nozzle was varied within the range of 1.5 l/min to 3.5 l/min, i.e. 0.0405 g/s $\leq G_2 \leq$ 0.0945 g/s. The channel section averaged values of temperatures of the internal and external flows of argon $T_1(z)$ and $T_2(z)$ were determined with allowance for heating the gas by the cathode surface (see Item 3.2.4). The initial temperature of the plasma gas was assumed to be equal to the temperature of water used to cool the cathode: $T_1(-L_w) = T_2(-L_w) = T_w = 300$ K. To calculate the heat exchange coefficients α_{c1} and α_{c2} which are involved in (3.59) and (3.60), the following criterion dependencies [142] were used:

$$\alpha_{ck} = Nu_k \frac{\chi(T_k)}{D_k} \left[\frac{\eta(T_k)}{\eta(T^c)} \right]^{0.14}, \quad (k = 1, 2),$$

(3.86)

where: $Nu_1 \approx 4$, $Nu_2 \approx 5$ are the Nusselt numbers for the internal and external flows; D_1, D_2 are the inside and outside diameters of the cathode, while the temperature dependencies

---

* The simplified LTE-model of the near-cathode plasma was used for the calculations.

of the transport coefficients χ and η for argon were calculated from formulae (2.46) and (2.47).

Results of numerical modelling of operation of the thermionic cathode under investigation, both in conventional arc mode and in the mode with laser heating of its tip, are given in Figures 3.5 to 3.9. In particular, Figure 3.5 shows distributions of temperature, current density and heat flow introduced by the arc over the working surface of the cathode without laser heating. It follows from Figures 3.5a and b that an increase in the arc current is accompanied by an increase in the temperature of the cathode surface and area of the cathode fixation of the arc, the increase in this area at $I > 200$ A occurring to a larger degree due to the cathode tip. This is associated with the fact that further immersion of the arc into the outlet hole of the cathode becomes disadvantageous in terms of energy, since it is accompanied by a considerable growth of the voltage drop in the region of plasma inside this hole.

Increasing the arc current from 100 A to 300 A causes the average current density at the cathode $\langle j_c \rangle = I / S_c$, where $S_c = \pi (R_c^2 + 2R_1 Z_c - R_1^2)$ is the cathode spot area for the tubular cathode (see Figure 3.2), to grow from 19.2 A/mm^2 to 29.2 A/mm^2. As to the heat flow averaged over the arc fixation area to such a cathode, $\langle q_c^a \rangle = Q_c^a / S_c$ where Q_c^a is the

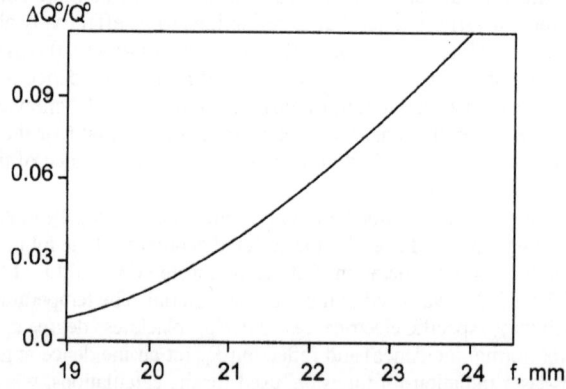

Figure 3.4 Part of the laser beam power, which is incident on the internal surface of the tubular cathode, as a function of its position with respect to the beam focus (TEM$_{20}$– mode; $\vartheta_b = 0.053$; $R_1 = 1.0$ mm).

Table 3.1 Transport, electrical and optical properties of tungsten.

T (K)	χ_c (W/(m K))	$\rho_c \times 10^8$ ($\Omega\,m$)	ε_c	Γ_ω	R_ω
300	130.0	6.0	0.03	0.019	0.980
600	126.0	9.2	0.05	0.028	0.970
900	121.0	17.6	0.09	0.050	0.947
1200	115.0	28.6	0.14	0.070	0.925
1500	110.0	40.7	0.19	0.084	0.910
1800	105.0	52.5	0.23	0.093	0.900
2100	101.0	65.1	0.27	0.099	0.894
2400	96.0	77.5	0.30	0.103	0.890
2700	92.0	91.1	0.33	0.106	0.887
3000	90.0	105.6	0.34	0.108	0.885
3300	89.0	118.1	0.35	0.110	0.883

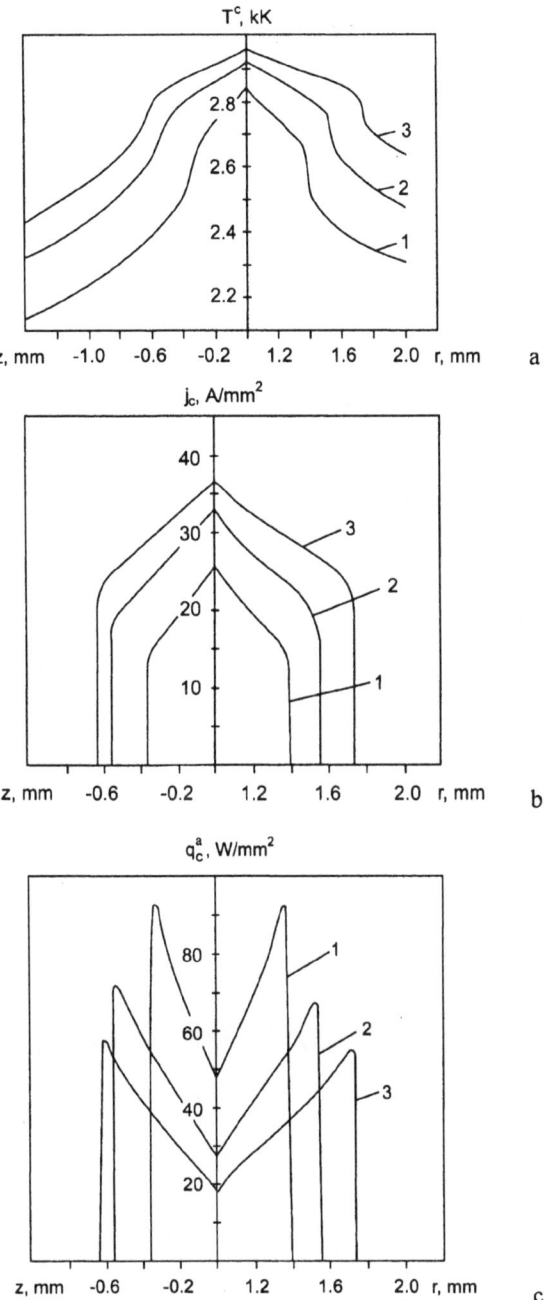

Figure 3.5 Distributions of the cathode temperature (a), current density (b) and heat flow (c) over the active surface of the thermionic tubular cathode in arc operation mode ($Q^0 = 0$ or $F < 17$ mm; $G_2 = 0.0676$ g/s) for different arc currents: 1, $I = 100$ A; 2, 200 A; 3, 300 A.

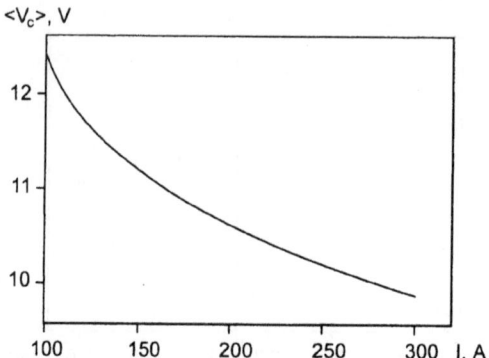

Figure 3.6 Averaged cathode drop for the thermionic tubular cathode in arc operation mode as a function of the arc current (parameters are the same as in Figure 3.5).

arc heat input into the cathode, calculated by integrating q_c^a for the cathode spot area, it decreases from 70.9 W/mm² ($I = 100$ A) to 37.8 W/mm² ($I = 300$ A). In addition, unlike the distributions of T^c and j_c having maximum values at points $r = R_1$; $z = 0$ (see Figures 3.5a and b), the maximum value of the heat flow q_c^a is found on the periphery of the cathode spot (see Figure 3.5c). It should be noted here that a similar distribution of heat flow at the cathode holds also for pin (non-tubular) thermionic cathodes [134].

Figure 3.6 shows the arc current dependence of the average cathode drop $\langle V_c \rangle$ calculated from equation (3.65), which uses values of components of the current density at the cathode (3.68) averaged over the area of the cathode fixation of the arc, the averaged cathode surface temperature ensuring the given density of the emission current and averaged plasma temperature T_c in the ionisation layer. The latter was found using the equation of balance of the near-cathode plasma energy (3.78), written with regard to the corresponding averaged values. It follows from the given calculated dependence that $\langle V_c \rangle$ for the tubular cathode investigated decreases with increase in I, which is generally characteristic of thermionic cathodes [119].

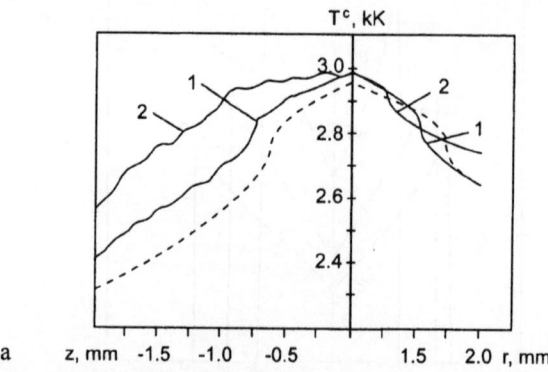

Figure 3.7 Distributions of the cathode temperature (a), current density (b), arc (c) and laser (d) heat flows over the active surface of the laser heated thermionic tubular cathode ($I = 300$ A; $Q^0 = 5$ kW; $G_2 = 0.0676$ g/s) for different distances from the cathode to the beam focus: 1, $F = 19$ mm ($Q_c^l = 166$ W); 2, 21 mm (325 W); dashed curves, $F < 17$ mm or $Q^0 = 0$.

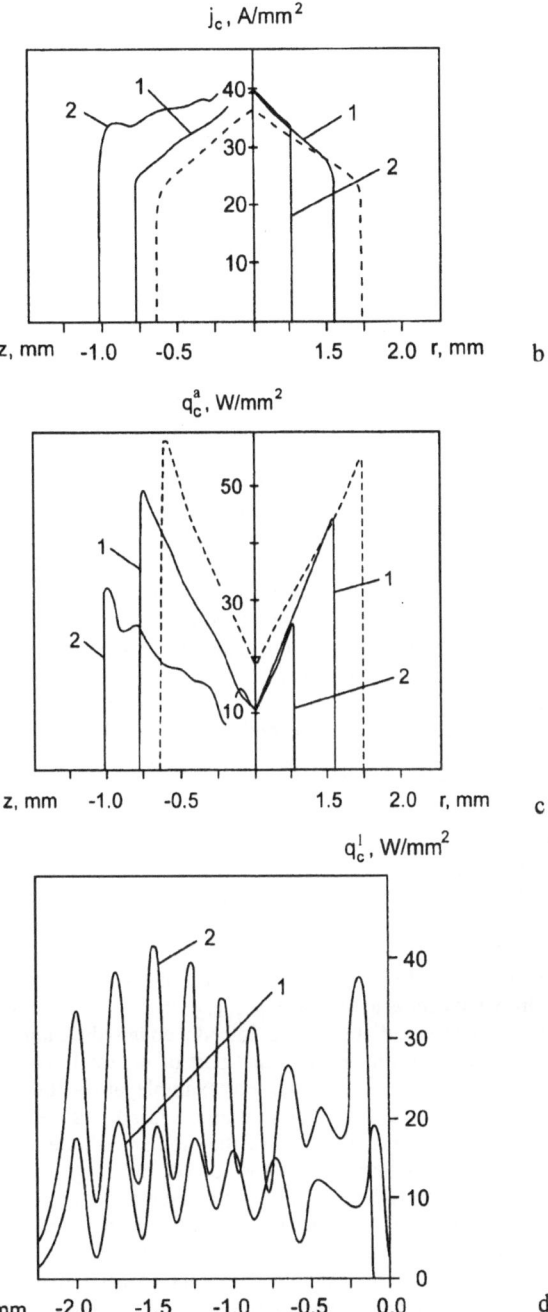

Figure 3.7 (*continued*).

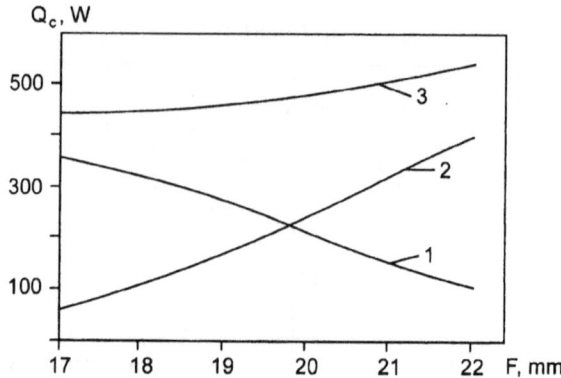

Figure 3.8 Components of the heat input for the laser heated thermionic tubular cathode as a function of its position with respect to the beam focus: 1, Q_c^a ; 2, Q_c^l; 3, Q_c^{tot} (parameters are the same as in Figure 3.7).

The effect of an additional laser heating the tip of the cathode under consideration on the character of occurrence of cathode phenomena can be illustrated using Figures 3.7 to 3.9. For example, Figure 3.7 shows distributions of the cathode temperature and the density of the current and heat flows over the working surface of the cathode, the latter being introduced, respectively, by the arc and laser radiation at different values of the laser heat input into the cathode Q_c^l, determined as the integral of q_c^l over the internal surface of the cathode. Growth of the laser heat input into the cathode taking place in the case of an increase in the distance of the focal plane of the beam $z = F$ from the exit section of the cathode $z = 0$ (see curve 2 in Figure 3.8), causes an increase in the surface temperature, which is especially pronounced inside the cathode outlet hole (see Figure 3.7a). The associated heterogeneous distributions of temperature and, hence, j_c and q_c^a (see Figure 3.7b and c) along the length of the outlet channel of the cathode is related to heterogeneous heating of its surface by laser radiation (see Figure 3.7d), which is enhanced by an increase in F (compare curves 1 and 2).

As far as the average current density at the cathode is concerned, as the cathode spot area decreases to some extent at increasing Q_c^l , and with constant arc current ($I = 300$ A), $\langle j_c \rangle$ grows from 29.2 A/mm² ($F < 17$ mm) to 37.4 A/mm² ($F = 22$ mm) and the cathode spot shifts from the tip of the cathode to its internal, more heated, surface (see Figure 3.7b). This immersion of the discharge into the cathode hole becomes possible owing to a decrease in the local cathode drop that occurs in this case (see, e.g., the behaviour of the average cathode drop $\langle V_c \rangle$ with increasing laser radiation power introduced into the cathode, shown in Figure 3.9) and leads to a fundamental increase in that part of the total discharge current which flows to the cathode within its outlet channel. For example, at $I = 300$ A, $Q^0 = 5$ kW and $F = 21$ mm ($Q_c^l = 325$ W), this part of the current is about 230 A and the current density in the plasma at the exit section of the outlet hole of the tubular cathode amounts to about 10^2 A/mm², which is much higher than that on the surface of the given thermionic cathode. Therefore, by using the additional laser heating of the internal surface of the tubular cathode, it is possible to effectively control the distribution of the current density and, hence, other parameters of the combined discharge plasma near this cathode.

An interesting peculiarity of the cathode phenomena for the given thermionic cathode preheated by laser radiation is the decrease in the heat flow introduced into this cathode by the arc at increasing Q_c^l and constant I (see Figure 3.7c). The average heat flow at the cathode $\langle q_c^a \rangle$ varies in this case from 37.8 W/mm² ($F < 17$ mm) to 13.4 W/mm² ($F = 22$ mm), while

Q_c^a decreases in such a way that the total heat input into the cathode $Q_c^{tot} = Q_c^a + Q_c^l + Q_c^j$, does not increase with growth in F to any significant extent (see Figure 3.8). It should be noted here that the power released in the bulk of the cathode due to the Joule heating Q_c^j does not exceed 10 % of Q_c^{tot}.

Therefore, the results of numerical modelling of the cathode phenomena for the suggested design of thermionic cathode of laser-arc plasma torch are indicative of very wide capabilities for controlling the processes that occur at the cathode and in the near-cathode plasma of the combined discharge due to additionally regulated heating of the working surface of this cathode by the laser beam that passes through it. This method of heating the cathode makes it possible not only to control characteristics of the plasma in the initial region of the laser-arc discharge during operation of the integrated plasma torch, but also to ensure a preliminary laser heating of the cathode to temperatures which provide minimum starting erosion at the arc ignition, which in turn results in a significant extension of life of this cathode.

The next stage in modelling the plasma torch designed for laser-plasma powder deposition consisted in a detailed numerical investigation into the combined discharge formed by this device, which disregarded feeding the additive powder into the discharge. The following dimensions of nozzles 3 and 4 and parameters which determine their locations relative to the cathode of the plasma torch investigated (see Figure 3.3) were selected: the radius of the cylindrical channel of the plasma-shaping nozzle R_{C1} = 3.0 mm; the length of this channel is equal to 4.5 mm; the inclination angle of the generating line of the internal conical surface of the nozzle is equal to 40 degrees; L_{C1} = 5.5 mm; the radius of the cylindrical channel of the powder focusing nozzle R_{C2} = 5.0 mm; the length of this channel is equal to 2.5 mm; the inclination angle of the generating line of the internal conical surface of the nozzle is equal to 45 degrees; L_{C2} = 10.0 mm. The flow rate of the plasma gas was set so that G_1 = 0.0135 g/s; G_2 = 0.0675 g/s (0.5 l/min and 2.5 l/min, respectively, at standard conditions); the volumetric flow rate of the carrier gas was assumed to be equal to 5.0 l/min, and its initial temperature to be equal to the temperature of the water-cooled walls of the channels, T_C = 300 K. Finally, the length of the open region of the discharge under consideration (distance from the exit section of the plasma torch to the anode surface d) was assumed to vary from 4 to 12 mm.

The characteristics of the plasma and the laser beam interacting with it were calculated on the basis of the LTE-model of the laser-arc discharge in a laminar gas flow (see Item

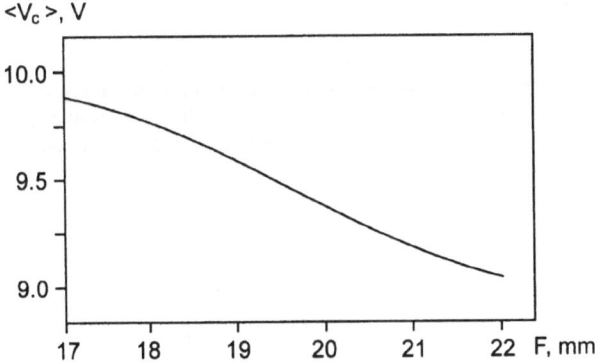

Figure 3.9 Averaged cathode drop for a laser heated thermionic tubular cathode as a function of its position with respect to the beam focus (parameters are the same as in Figure 3.7).

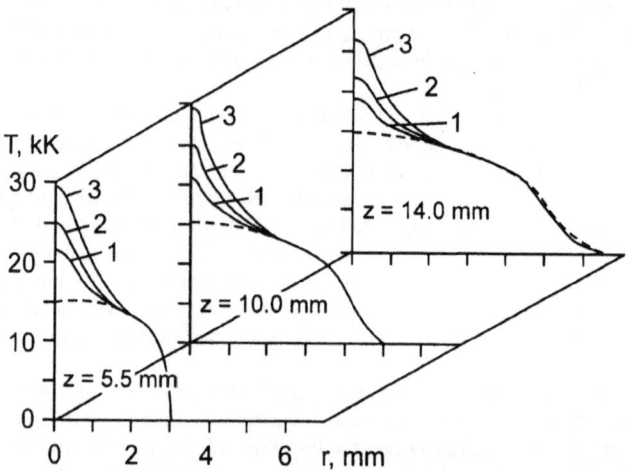

Figure 3.10 Spatial distributions of the temperature of plasma generated by the laser-arc plasma torch ($I = 200$ A; $d = 12$ mm; $F = 14$ mm) for different laser beam powers: 1, $Q^0 = 1$ kW; 2, 2 kW; 3, 4 kW; dashed curves, $Q^0 = 0$.

3.2.1). The above calculation results for the distribution of the current density j_c and temperature T_c of the plasma in the region of the cathode fixation of the arc were used to assign boundary conditions near the surface of the tubular cathode of the plasma torch. To exclude the anode phenomena from consideration, the length of the calculation domain of the discharge L was chosen to be somewhat smaller than the distance from the cathode to the anode. This choice enabled the "soft" boundary conditions (3.25) to be used as the boundary conditions for $z = L$. In addition, the calculations did not allow for the contribution of the laser radiation reflected from the anode surface into the energy balance of the plasma, because it turned out to be small for the chosen conditions of realisation of the combined discharge.

Some results of numerical modelling of the laser-arc discharge for different modes of operation of the plasma torch investigated are shown in Figures 3.10–3.15. In particular, Figure 3.10 shows spatial distributions of the plasma temperature for the conventional transferred arc (dashed curves) and for the same arc but affected by the focused laser beam (solid curves). As it follows from the above curves, absorption of the laser radiation by the arc plasma leads to a substantial increase in temperature in its central region, the maximum possible values of T being increased with an increase in the beam power Q^0. In turn, this growth of the plasma temperature leads to an increase in its electrical conductivity (see, e.g., Figure 2.6) and, hence, to an increase in the current density in the near-axis zone of the discharge (see Figure 3.11). Therefore, the laser-arc discharge generated by the integrated plasma torch is characterised by an increased concentration of the thermal and electrical energy in the laser beam affected region of the plasma and by a high spatial stability of this region, which is strongly related to the beam axis.

The described variation in the thermal mode of burning of the transferred arc under the laser radiation effect causes a fundamental redistribution of gas-dynamic characteristics of the flow of plasma generated in the laser-arc plasma torch with an increase in Q^0. One of the main causes of this phenomenon is the decreasing viscosity of the argon plasma observed with an increase in its temperature (see, e.g., Figure 2.4). Another cause of this phenomenon is the above-mentioned redistribution of the current density in the discharge, which enhances

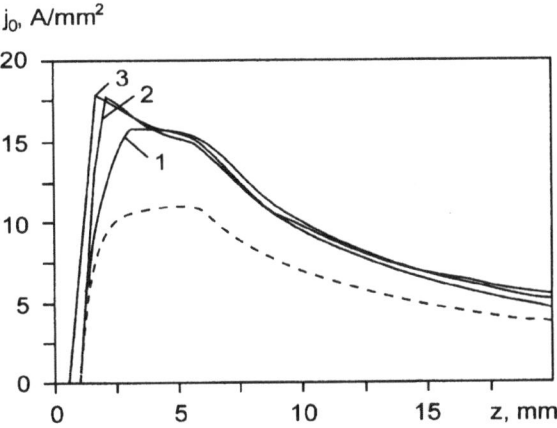

Figure 3.11 Distributions of the axial current density along the combined discharge axis (parameters and designations are the same as in Figure 3.10).

the role of electromagnetic forces in acceleration of the plasma flow. This results, as it is shown in Figure 3.12, in a marked increase in the axial component of the velocity of the plasma near the discharge axis. Despite this increase in the velocity of the plasma, a decrease in its density with an elevation of temperature (see, e.g., Figure 2.2) leads some decrease in the gas-dynamic pressure of the plasma flow $\rho u^2/2$ in the near-axis zone of the combined discharge (see Figure 3.13). It should be noted that the decrease in $\rho u^2/2$, which causes a decrease in the dynamic effect on the molten metal surface, is a very important feature of the process of surfacing involving the integrated plasma torch [143].

Interaction of laser radiation with arc plasma occurring in the device considered leads not only to the redistribution of characteristics of the latter, but also to changes in the laser beam properties, caused by its absorption and refraction in the plasma. For example, absorption of laser

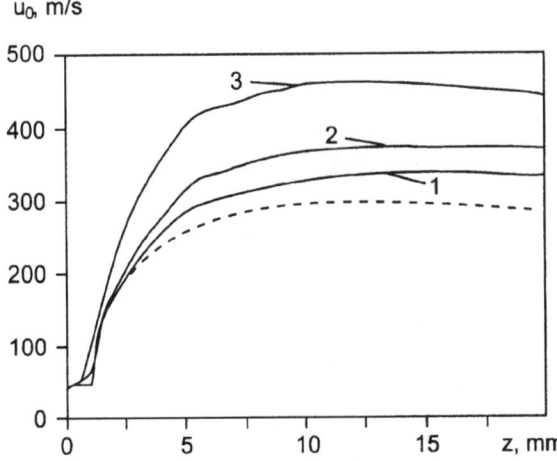

Figure 3.12 Distributions of the axial velocity of plasma generated by the integrated plasma torch along the discharge axis (parameters and designations are the same as in Figure 3.10).

radiation by the plasma leads to the fact that at $z = 20$ mm the beam power for the conditions under consideration constitutes only about 30 % of Q^0. The associated distribution of axial values of the radiation intensity along the discharge at $I = 200$ A are shown in Figure 3.14. As it follows from the curves shown in the Figure, propagation of the laser beam in a heterogeneous plasma causes its additional focusing (compare corresponding solid and dashed curves), which is enhanced with growth of Q^0. Figure 3.15 illustrates the effect of the arc current on the focusing properties of the combined discharge plasma. In general, the calculated data given in Figures 3.14 and 3.15 indicate that focusing the laser beam in the laser- arc discharge generated by the suggested plasma torch can be effectively controlled through varying the arc current and power of this beam.

The results of computer simulation of the phenomena occurring at the cathode and in the column of the discharge investigated were utilised for development of a specialised integrated plasma torch for laser-plasma powder deposition (see Figure 3.3) and for selection of optimal conditions for its operation. This device was designed and its prototype was manufactured by the Plasma-Master Co. Ltd. (Ukraine) in 1994. The main peculiarities of this plasma torch prototype design are as follows: tubular water-cooled tungsten cathode; adjustment of the cathode position with respect to the laser beam focal plane; gas seal for protecting the laser beam focusing lens; internal coaxial system for a distributed feed of the additive powder.

The experimental investigation of the laser-arc plasma torch prototype was carried out at the Fraunhofer-Institut fuer Produktionstechnologie (Germany), using a RS-5000 CO_2-laser and a plasma arc power supply Messer Griesheim Unitig GW 30. An external view of the experimental arrangement is shown in Figure 3.16. The target of the experimental investigation was to study peculiarities of operation of the integrated plasma torch without feed of the additive material, identify the prospects for using this device for laser-plasma powder deposition and define the basic technological capabilities of the above process. In all the experiments the arc current was varied from 100 A to 280 A, the power of the laser beam was varied from 0 to 4 kW, the position of the focal plane of the initial beam with respect to the cathode, F, ranged from 14 mm to 22 mm, the distance from the exit section

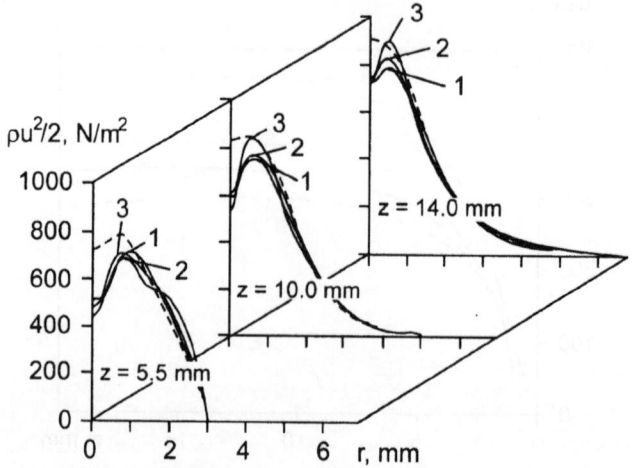

Figure 3.13 Spatial distributions of the gas dynamic pressure of plasma flow generated by the laser-arc plasma torch (parameters and designations are the same as in Figure 3.10).

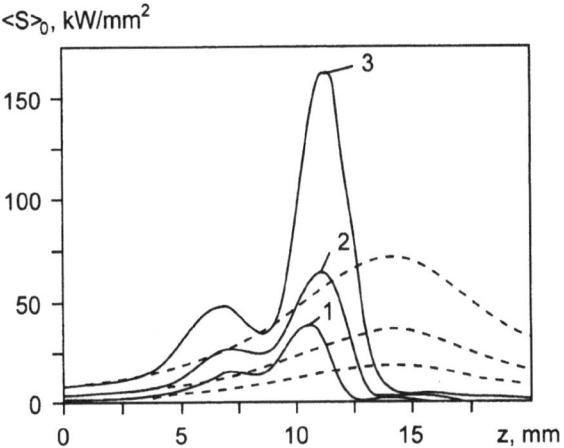

Figure 3.14 Distributions of the laser radiation intensity along the combined discharge axis for different laser beam powers (parameters and designations are the same as in Figure 3.10).

of the plasma torch to the anode (workpiece) surface *d* was selected to be equal to 4, 8 or 12 mm, and flow rates of the plasma and carrier gases were kept constant and equal to the values indicated above.

The experiments conducted showed that within the entire range of the arc currents and power of the laser beam investigated the laser-arc plasma torch exhibited stable operation and a high space-time stability of parameters of the generated plasma. Voltage-current characteristics of the combined discharge with a copper water-cooled anode were measured at different values of Q^0, F and d. It was established that for the selected range of operational

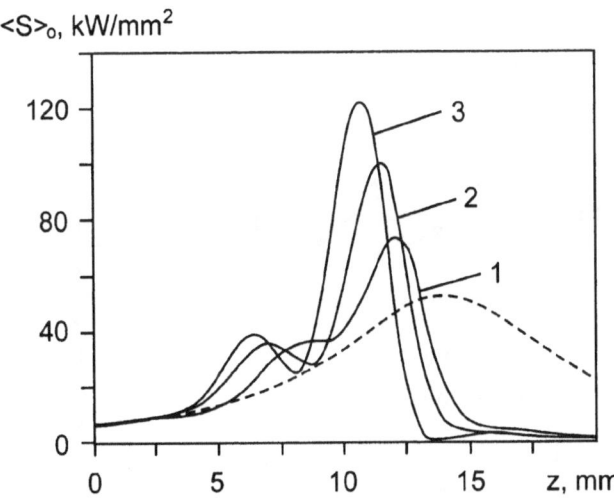

Figure 3.15 Distributions of the laser radiation intensity along the combined discharge axis ($Q^0 = 3$ kW; $d = 12$ mm; $F = 14$ mm) for different arc currents: 1, $I = 100$ A; 2, 200 A; 3, 300 A; dashed curves, $Q^0 = 0$.

Figure 3.16 An external view of the integrated plasma torch prototype at the experimental arrangement for laser-arc powder deposition.

parameters of the plasma torch the discharge investigated had a rising voltage-current characteristic, and that the higher the beam power, the more marked was the growth of voltage with increase in the arc current or discharge length (distance *d*). In addition, the measurements proved the theoretical conclusion that interaction of laser radiation with arc plasma leads to a decrease in voltage of the combined discharge, and that this decrease, as it follows from the experimental data shown in Figure 3.17 (see solid curves), is most pronounced at a laser beam power of $Q^0 < 2.5$ kW. The calculated dependence of the total voltage of the discharge upon the beam power, shown in this Figure (see dashed curve), demonstrates an acceptable agreement between the experiment and the calculated values of *U* determined as a sum of the cathode voltage drop, voltage drop at the plasma column and

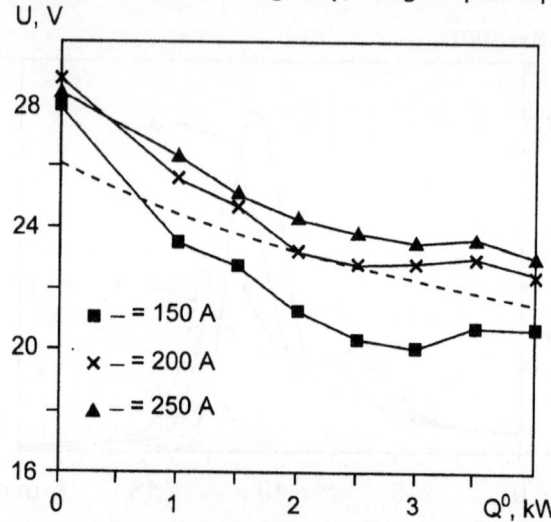

Figure 3.17 Dependences of the combined discharge voltage ($d = 8$ mm; $F = 16$ mm) on the laser beam power for different arc currents: solid curves, experimental data; dashed curve, calculated data for $I = 200$ A.

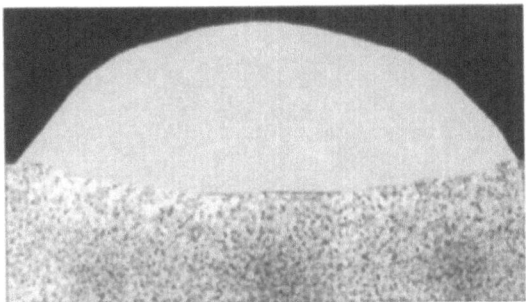

Figure 3.18 Macrosection of a single bead deposited by the laser-plasma surfacing process ($I = 200$ A; $Q^0 = 3$ kW; $d = 8$ mm; $F = 14$ mm; cladding speed is 18 m/h; powder rate is 3 kg/h).

the anode voltage drop, V_a, which was estimated for the case of the non-evaporating anode on the basis of the following expression [120]:

$$j_a = \frac{1}{4} e n_e \sqrt{8 k_B T_a / \pi m_e}\ \exp\left(\frac{e V_a}{k_B T_a}\right) - e \sum_{\alpha = 3, 4} Z_\alpha n_\alpha \exp\left(\frac{Z_\alpha}{2}\right) \sqrt{k_B T_a / M}\,. \tag{3.87}$$

Here: j_a and T_a are the current density and temperature of the plasma near the anode, which were selected, e.g., to be equal to the calculated values at the discharge axis at $z = L$; n_e, n_3 and n_4 are the equilibrium concentrations of electrons of singly and doubly charged ions in the plasma, corresponding to the above temperature; and V_a is the anode potential drop to be determined ($V_a \leq 0$). To complete the description of the results of the experimental investigations of the operation of the integrated plasma torch without the feed of an additive powder, it should be noted that the described character of the effect of the laser radiation on the combined discharge voltage depends hardly at all on the position of the focus of the beam used (distance F).

The operational capabilities of this plasma torch were investigated by an example of laser-plasma surfacing flat samples of low-carbon steel 20 mm thick. HASTEL-LOY C and Stellite 6 powder alloys with particles 20–63 µm in size were used as the additive material. The investigations show that the combination of the plasma arc with the laser beam in the integrated plasma torch enables the surfacing speed to be increased 2–3 times, as compared with the conventional method of plasma-powder deposition (PTA process) at the same plasma arc current, owing to improvement in the spatial stability of the discharge. For example, single beads with a good formation could be deposited at a speed of up to 50 m/h. In addition, due to a decrease in the dynamic effect of the plasma flow on the surface of a melt, which occurs under the influence of the laser radiation, under the optimal conditions of laser-plasma surfacing the penetration of the base metal did not exceed 10 %, as is shown, for example, in Figure 3.18.

In general, the results of computer simulation and experimental investigations of the integrated plasma torch for laser-plasma powder deposition lead to the conclusion that the application of such a device for surfacing is highly efficient from the viewpoint of better productivity, stability and reproducibility of this technological process. In addition, the developed laser-plasma torch can be successfully utilised for welding and other related processes of metal treatment. Same examples of the present practical application of various laser-arc and laser-plasma processes will be considered in the next chapter.

CHAPTER 4

Practical Applications of Laser-Arc Processes

4.1 Hybrid and Serial (Combined) Methods — Two Kinds of Laser-Arc Processes

If laser beam welding (LBW) is combined with the arc welding method, the procedure versions shown in Figure 4.1 can always be differentiated. During serial (combined) processing the laser beam and electric arc work locally and temporally separated from each other. This technique is applied to single-sided welding of webs at a given laser power as a continuous tack weld with a subsequent Gas Metal Arc (GMA) or submerged arc welding process. This can be applied in the field of manufacture and laying down of large pipes and in shipbuilding with weld preparation as required for arc welding. The increased penetration depth of the laser beam allows thicker webs to be welded with lower welding deformations than is possible or common with GMA welding.

Welding of a root pass, which is required in many cases, but has to be reworked in a time consuming process afterwards, can be avoided in the case of a laser welded thick web. With submerged arc welding, the grooving-out required for high webs prior to welding of the cap run is not necessary. Due to the reduced groove cross section there is also a decrease in the required quantity of filler metal, in the total heat input and in the production time. Compared to conventional techniques, especially in the production of pipes with inner plating considerable quantities of expensive, high-alloyed filler metals can be saved by the application of laser to weld thick webs.

On condition that sufficient quality of the weld root can be attained there is the possibility of one-sided welding without complex support of the molten pool. The heat subsequent to the GMA welding process can cause an additional limited effect of tempering — the so-called short-time post-heating of the mostly hard structure of the laser beam welded seam.

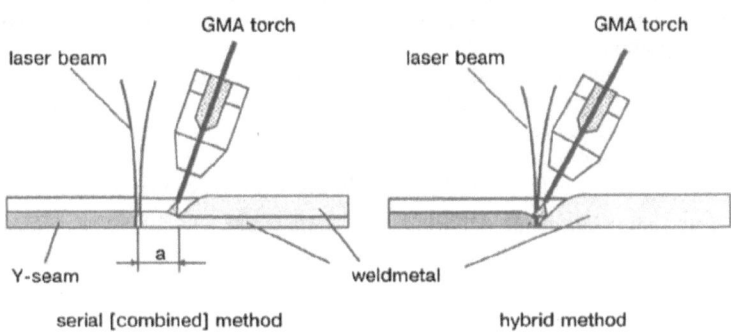

Figure 4.1 Two kinds of combination of laser and arc welding processes.

4.2 Laser + Gas Tungsten Arc Welding

4.2.1 Feasibility of laser + GTA welding

One of the initial works on combined welding technique was performed by Steen and Eboo [6]. They demonstrated that high speed, low current Gas Tungsten Arc (GTA) welding of mild steel is stabilised by the presence of the laser beam. Further work by Diebold and Albright [41] demonstrated acceptable welds in aluminium alloys achieved owing to high welding speeds. Laser operation has been done with no more than 600 W CO_2-laser output power.

Results so far have shown that the combined action of Nd:YAG-laser beam and GTA (TIG arc) has an apparent synergistic effect [144], producing a larger volume of the molten weld pool than those of the two processes separately for a given welding speed, Figure 4.2.

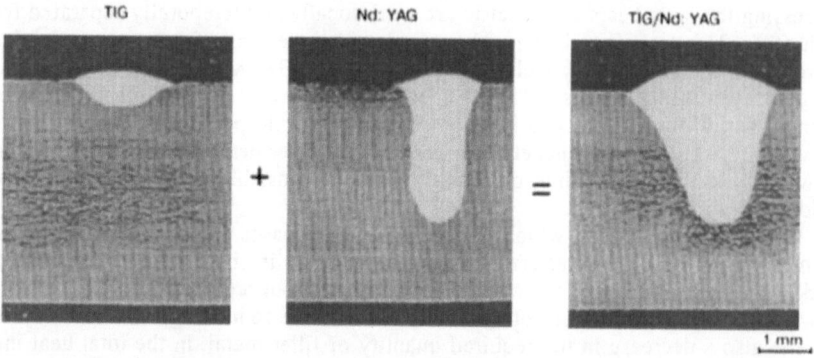

Figure 4.2 Weld section geometry for TIG, Nd:YAG and hybrid TIG/Nd:YAG welding (material: 304, welding speed 1 m/min, laser power 2.3 kW (CW), focal distance 100 mm, focused beam radius 400 μm) [144].

Despite these potentially significant gains resulting from TIG augmentation of laser beam welding* it is probable that the TIG process is connected with a number of inherent limitations, including:

1 Erratic and inconsistent arc ignition due to the use of a high frequency, high voltage current. Repeated use of high frequency arc ignition causes erosion of the electrode point and destabilisation the arc.
2 Lack of stiffness and directionality of the free burning arc. The arc will seek the nearest earth point, which may not be the most effective or stable position for the process.
3 The projecting tungsten electrode is exposed to hot metal vapours causing contamination and rapidly deteriorating arc stability. High affinity of tungsten for oxygen can cause the electrode to oxidise quickly, causing the arc to become erratic [30].

It is also a question of the material to be welded whether the result of welding is successful. Especially in the case of aluminium welds there are a lot of difficulties, such as:

---

* In classification of the hybrid processes suggested in Chapter 1 this process can be referred to as laser + GTA process.

1 Aluminium alloys have a higher reflectivity to CO_2-laser beam than ordinary mild steel. Absorption of the laser beam into solid aluminium is not only low but sometimes erratic, and the reflected laser beam may cause undesirable damage to optics. Also in the case of absorption in "key-hole" the limitations of the TIG process are still working.
2 Bead formation in welding of some aluminium alloys is irregular, and sometimes heavy sputtering occurs. A rough surface and deep undercuts can be formed.
3 Large blowholes or cavities and weld cracks were found in the welding of some aluminium alloys.

The idea of combining LBW with arc welding is not new and was tried by several researchers in the past (see, e.g., [145]). But in the case of aluminium welding most of the researchers could not find a favourable using of combination of laser beam and GTA welding, because the tungsten electrode was severely damaged in a short time and many weld cracks were found in the longitudinal sections of weld beads [146]. In the case of steel welding there are some improvements provided by combined LBW and GTA welding, in comparison to aluminium welding.

4.2.2 Principles, devices and practical examples of laser + GTA welding

In the case of serial (combined) laser + GTA (laser TIG) welding there are two kinds of torch positions.

Laser beam welding with pre-running TIG arc

With the arc arranged ahead of the laser beam molten pool, as shown in Figure 4.3, a pre-heating effect is obtained. At currents required for pre-heating, melting of the sheet surface must surely be avoided. Therefore, the arc length must be above 10 mm. Although this reduces the efficiency of the process, the net heat input into the workpiece is kept almost constant. In our experiments the electrode spacing was 14 mm [147]. Under these conditions a stable arc can be maintained only with argon shielding. It is not possible to have such lengths of the arc with helium shielding. Another parameter influencing the pre-heating effect of the TIG arc is the distance between the TIG torch and laser beam in the arrangement. The distance between the two heat sources must be kept at minimum, because charge carriers

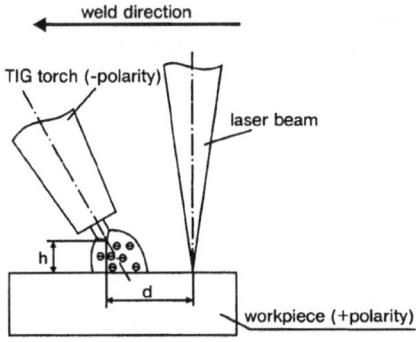

Figure 4.3 Arrangement of laser beam and TIG arc (h – electrode spacing, d – distance between TIG torch and laser beam).

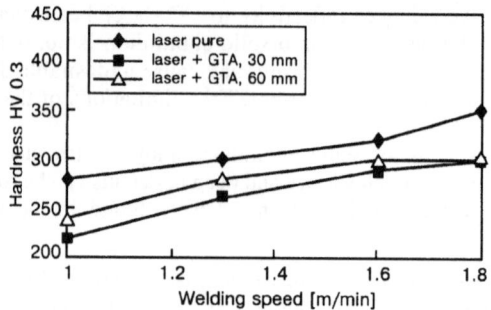

Figure 4.4 Hardness values in the welds measured in the weld metal at half of the penetration depth (hardness in the lower half of the weld for all welds 400 HV, HV_{max} in the weld bottom)

of the arc getting into the laser beam would disturb the laser beam welding process due to absorption and would finally break down the deep penetration effect. The effect of the distance between the laser beam and TIG arc on the superposition of the temperature fields depends on the welding speed.

The influence of pre-heating on the weld width and the weld shape can be demonstrated. At a laser beam to TIG torch distance of 30 mm a widening of the weld top bead and heat-affected zone is obtained in contrast to pure laser welding. At a distance of 60 mm this effect reduces and can hardly be noticed. The influence on the weld width on the top side must be considered to be a positive phenomenon, because the weld shape represents an effective means to avoid solidification cracking in the sense of a continuous solidification of the molten pool from the bottom of the weld to the top side. Significant changes in the weld shape and solidification conditions may only be expected with a higher heat input (> 2 kW). For laser beam welding the penetration is first of all characterised by the laser beam parameters, especially by the power density. The influence of the TIG pre-heating on the penetration depth can be neglected.

Apart from the positive influence on the weld width there is also a positive effect on the reduction of hardness of the weld. This reduction of hardness is shown in Figure 4.4 for steel St37. The smaller the distance between laser beam and TIG torch, the larger the reduction.

In the junction of *I*-welds having a thickness of 12 mm a formation of solidification cracks in the centre of the weld can not be avoided by TIG pre-heating at welding speeds of 1 and 1.3 m/min. It is not possible to increase the welding speed by a maximum focusing of the laser beam on the sheet surface even with a limited pre-heating with the TIG arc. In order to control the weld shape focal positions in the range of −2 to −6 mm are required, which would reduce the welding speed to 0.8 m/min with the *I*-welds of 12 mm thickness.

Laser beam welding with following TIG arc

The use of an arc following the laser beam has the aim to obtain a rounding of the back side of the molten pool by an additional heat input and in this way a change in the spatial orientation of the dendrite growth occurs, according to Figure 4.5a, as compared with pure LBW (Figure 4.5b), where the weld shape is favourable for formation of solidification cracks.

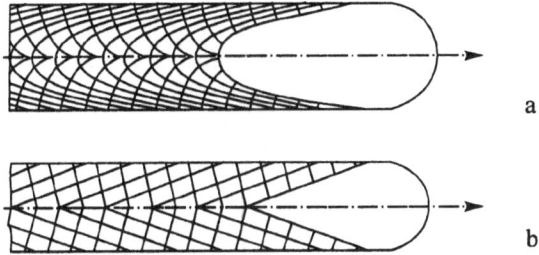

Figure 4.5 Influence of welding method on the solidification of the weld metal: (a) laser beam welding with following TIG arc; (b) laser beam welding.

The welding process in this arrangement for single-pass welds (Figure 4.6) seems to be very critical, because the distance between the laser beam and arc must be kept necessarily such that the arc burns on the rear end of the molten pool. There is the risk that charge carriers from the arc may come into the laser beam and lead to dramatic effects of absorption and scattering.

A sheet of 17 mm thickness is taken as an example to show that at lower welding speeds a weld formed has the shape, which corresponds to the intentions of the combined welding method. The weld has the shape of a wineglass having a maximal weld width on its top, which would not have been achieved by the laser alone, Figure 4.6b. The melt starts solidifying from the bottom. So the melt can always flow into hollow spaces that might be formed and thus cure solidification cracks. This weld shape must therefore be considered as a variant highly resistant against hot cracking. However, the attained welding speeds are clearly below 1 m/min.

In conclusion to this item we should note interesting results obtained at TU Braunschweig (Germany) with laser TIG welding of steels and aluminium alloys [148]. In

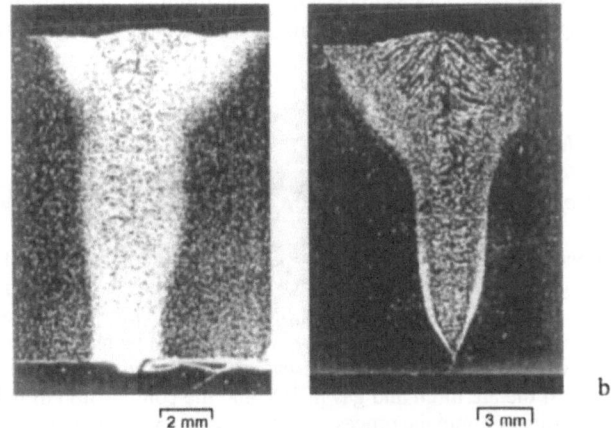

Figure 4.6 Single-pass hybrid welding with following TIG arc: (a) 12 mm sheet thickness at 1 m/min, insecure process, insufficient efficiency; (b) 17 mm sheet thickness at 0.6 m/min, secure process, positive efficiency (macrosections, etching: nital).

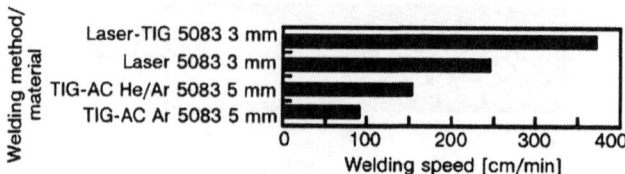

Figure 4.7 Comparison of welding speeds for TIG, laser beam and laser-TIG processes at optimal weld profiles for sheet thickness 3 mm and 5 mm (5083 – AlMg4.5Mn0.7) [148].

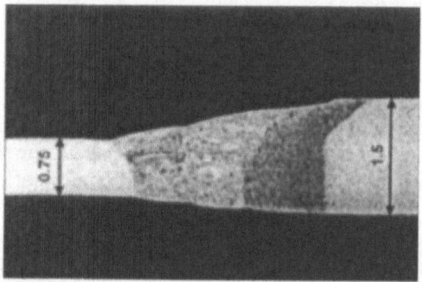

Figure 4.8 Laser TIG welding of steels DC05 (0.75 mm) and Z StE340 (1.5 mm) with organic coatings on both sides (Nd:YAG-laser beam power 3 kW, TIG arc current 200A, welding speed 5.0 m/min) [148].

particular, Figure 4.7 shows data on the speed of welding aluminium alloy AlMg4.5Mn0.7 using different welding methods. Figure 4.8 shows a cross-section of the welded joint between steels of different thickness with organic coatings deposited on both sides. The data given prove that hybrid laser TIG welding holds high promise for practical application in terms of increasing the welding speed and producing the high-quality permanent welded joints.

4.3 Laser + Gas Metal Arc Welding

4.3.1 Feasibility of laser + GMA welding

One very fundamental matter of hybrid laser + GMA welding is the main character of the process. Depending on the energy ratio of the two energy sources the character of the laser + GMA process may be either more arc-like or more laser-like. The process characteristic has an immediate effect on the weld geometry. The welding depth increases with an increase in the laser energy, whereas the weld width increases with the arc energy. Furthermore the geometrical set-up of the arc torch and gas nozzle and the composition of the working gas have an important influence onto the process. In most cases a preceding torch leads to a much more stable process. At the same time the range of interactions between the two processes increases. For this set-up the weld geometry was found smaller and the dilution of the filler material much more homogeneous in comparison with the laser preceding the set-up. Concerning the torch position, the best results were obtained with a torch in strongly inclined position, which means a very small angle between the torch and the laser axis. The distance

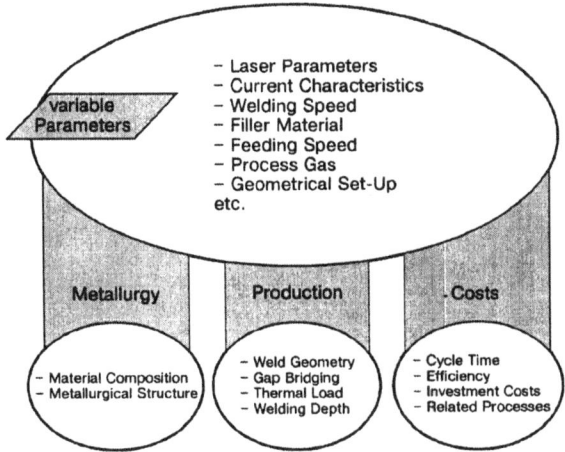

Figure 4.9 Variable parameters for hybrid laser + GMA process [149].

between the arc foot point and the laser beam has to be in-process adjusted, as it has a strong influence onto the process performance [149].

Different studies on the arc characteristics have shown that the composition of the gas has a strong effect on the arc and the dynamic behaviour of the molten material. This effect was observed for the hybrid process as well. Anyway, the working gas heavily affects the laser beam — material interactions in the LBW single process as well. Different working gas components may have different and even opposite effects in two processes. Although a low level of ionisation potential stabilises the arc process, it increases absorption of the laser beam in the laser-induced plasma. This means a decrease in the laser energy at the workpiece. The composition of the gas as well as the set-up of the gas nozzle are process critical and have to be optimised for each material and energy level, Figure 4.9.

The set-up with the task adjusted parameters demands a high level of knowledge and accuracy, though theoretical understanding as well as the practical experience for the process are still rather low. Estimation of the optimised parameters is not possible. As it has been observed, these parameters have a clearly pronounced effect on interactions of the single processes and their synergistic effects, but explanations for this correlation are still very weak. However, once the process has been worked out for a special task of joining a certain workpiece, it works very stably with reproducible results and high process security.

Energy aspects of the hybrid process

From the energy point of view, increase in the process efficiency is the most pronounced aspect of the hybrid process. It is a matter of fact that the sum of the energies of the two single processes is less than the energy of the hybrid process. In this respect the potential of this process is far beyond any other comparable process, Figure 4.10.

To explain this increase in the process efficiency two different effects have to be mentioned. On the one hand, a higher energy density leads to a higher welding speed. The loss of energy by convection into a workpiece is reduced. On the other hand, the process interactions lead to synergistic effects as mentioned above. These effects have been observed and de-

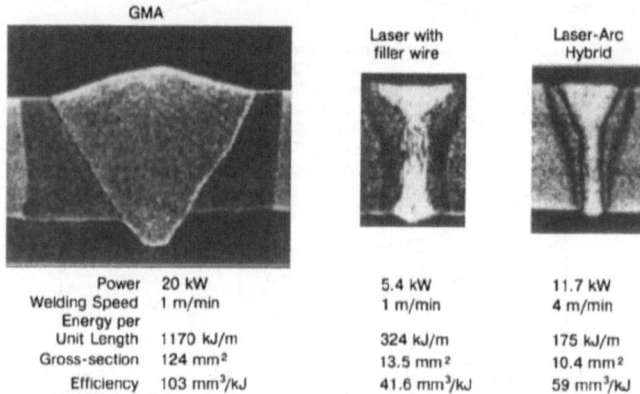

	GMA	Laser with filler wire	Laser-Arc Hybrid
Power	20 kW	5.4 kW	11.7 kW
Welding Speed	1 m/min	1 m/min	4 m/min
Energy per Unit Length	1170 kJ/m	324 kJ/m	175 kJ/m
Gross-section	124 mm²	13.5 mm²	10.4 mm²
Efficiency	103 mm³/kJ	41.6 mm³/kJ	59 mm³/kJ

Figure 4.10 Comparison of GMA, laser with filler wire and laser-arc hybrid welding processes (plate thicknes: 6 mm) [149].

scribed. It is to understand that for steel only the use of a CO_2-laser and the related generation of a laser induced plasma lead to stabilisation of the arc, which is attracted by the "key-hole". For aluminium alloys this effect does not depend on the kind of the laser [149]. The reason for the arc guidance may be removal of the aluminium oxide layer with low conductivity. However, the process combination leads to improved coupling of the laser energy. No complete self-sufficient description and explanation for the phenomenon can be given yet.

The efficiency of a welding process can best be expressed in terms of molten material per unit energy. Figure 4.10 gives a value of the used energy per unit volume of molten material with no regard to evaporation and spattering. The use of a filler wire for steel welding leads to reduction of the process efficiency. A decrease in the efficiency can be explained by the reflection of the beam at the filler wire, which occurs together with the plasma formation. The hybrid laser + GMA process is characterised by a high degree of variability, affecting especially the process efficiency. The resulting values are influenced by the welding speed, power and especially by the tuning of one process to the other. In the investigation of [149] the maximum observed value is about 70 mm³/kJ, with an equal amount of the laser and arc power. The tendency observed was toward an increase in the efficiency with a decrease in the energy per unit length. The minimum was about 40 mm³/kJ with a power share of the laser of 40%. The average value was about 50 mm³/kJ.

Enhanced welding possibilities by variation of the energy per unit length

The energy investigation of the hybrid laser + GMA process shows its high variability in terms of the energy and welding speed. In this context the energy that is used per unit length is the most important and dominating factor of the metallurgical aspects in welding. Although the process studied, used a filler wire, the minimum value was minor, as compared with that of the pure laser process without filler material [147, 149] and vice versa, very high values could be realised as well, although the weld geometry and characteristics are very laser-like.

The use of a filler wire is very sensitive to the laser beam process. In the arc augmented process it becomes much easier and safer. The wire is melted completely in the arc and brought into the process zone with a high process safety. The advantages of the use of a filler wire, such as enhanced gap bridging and possibility of metallurgical influence, can fully be used. The workpiece preparation is easier because of lower accuracy demands. This leads to cost reduction. Existing structures which are designed to be welded by arc processes are difficult to adapt to the laser beam process. The design of the joints has to be redone systematically. Due to the arc characteristics of the combined process, existing structures can easily be manufactured. Further, the coupling of the energy as well as the process efficiency are increased. Especially for materials that are difficult to weld the process is stabilised. The reduction of the energy per unit length leads to a decreased thermal load on a workpiece. The thermal distortion and the residual stress in the workpiece are reduced remarkably.

In the manufacturing process very often the main process is not a critical point. There are the preceding processes, such as workpiece preparation, clamping, tacking, and the following processes, such as the reset of the workpiece, that are critical and cost intensive. This is where the advantages of the hybrid laser + GMA process can have enormous effects. The higher working speed leads to lower cycle times, which are calculated directly per unit production costs. Finally, the reduction of the specific machine costs by the hybrid process are mentioned. For a given application with a defined welding speed and depth the laser power can be reduced. This leads to reduction in both investment and running costs without any significant change in the weld and process characteristics. The costs of the whole machine with the same production characteristics are much lower than for the laser beam welding machine.

4.3.2 Principles, devices and practical examples of laser + GMA welding

Recent investigations have focused on the application of high power lasers in combination with arc welding processes to weld plate thickness above 12 mm [147]. Laser + GMA welding was performed using a CO_2-laser with a maximum output power of 12 kW and a metal arc welding power supply rated up to 500 A direct current at the SLV Mecklenburg-Vorpommern (Germany) [150].

In the laser + GMA welding technique a laser beam is placed on the working point of a metal arc as shown in Figure 4.11. That means both processes are operating simultaneously, resulting in combined plasma results. The laser beam ensures joint penetration. The GMA process works as an additional heat source and supports the laser beam. Table 4.1 shows the cross-sections for the laser beam welding and laser + GMA welding processes.

The higher welding speed in the case of combined laser + GMA welding and the decreasing specific heat input at the same joint penetration should be noted. Beside these effects the combined welding technique allows a higher gap bridging, than that in the case of laser welding. The accuracy requirements for laser beam welding only are much higher. The expenditure for seam preparation in the case of the combined technique is much lower. The possibility for laser beam welding of long welds, as is typical in structural steel and shipbuilding is improved. Welding test samples meets the requirements of ships classification societies concerning tensile strength, yield strength, elongation, hardness and toughness as defined in "Guidelines for the approval of CO_2-laser welding" [151]. Figure 4.12 shows very clearly the improvement by hybrid laser + GMA welding for T- and butt joints. This figure shows for shipbuilding conditions that the welding speed is twice as high and the shrinkage is very small. Also the classification requirements are fulfilled in comparison to the pure laser welding process alone. At the same time, Table 4.2 shows a relatively small shrinkage in the case of hybrid welding in comparison to other kinds of welding processes. Nevertheless, there is a need to improve results in the case of hybrid

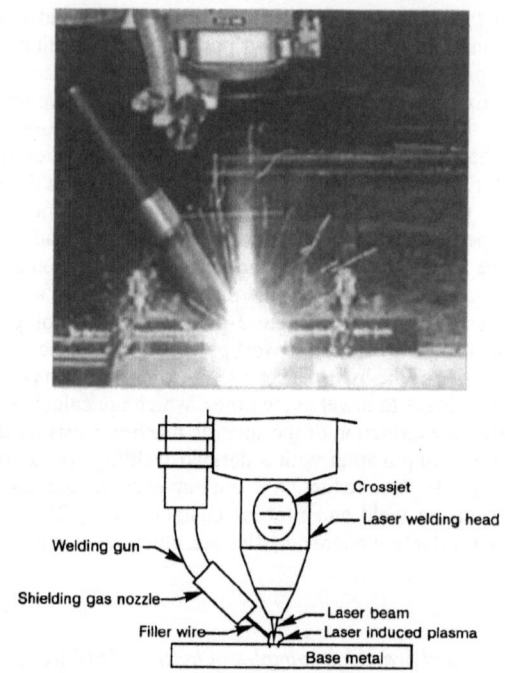

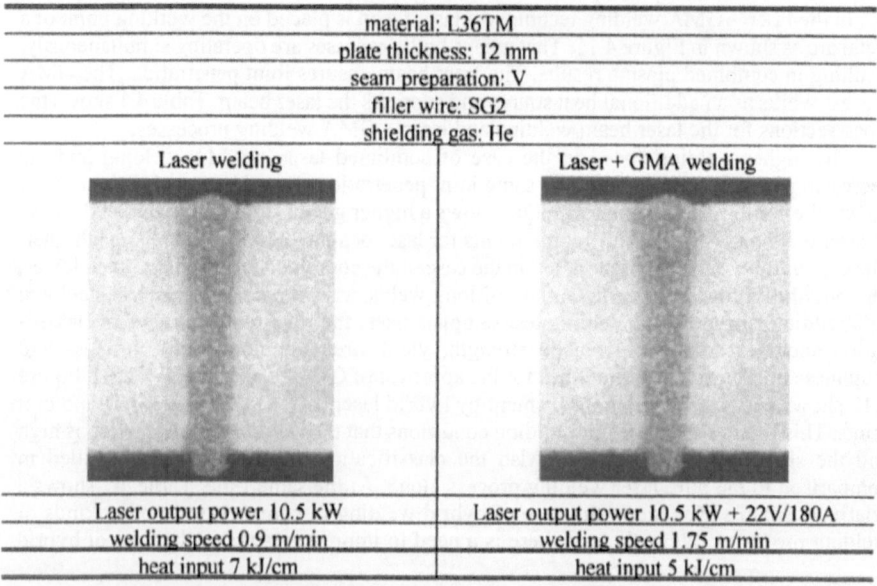

Figure 4.11 External view of laser-hybrid welding (a) [information courtesy of Fronius Schweissmaschinen KG (Austria)] and schematic diagram of the laser + GMA process (b).

Table 4.1 Comparison of weld cross-sections in laser welding and laser + GMA welding.

material: L36TM
plate thickness: 12 mm
seam preparation: V
filler wire: SG2
shielding gas: He

Laser welding	Laser + GMA welding
Laser output power 10.5 kW	Laser output power 10.5 kW + 22V/180A
welding speed 0.9 m/min	welding speed 1.75 m/min
heat input 7 kJ/cm	heat input 5 kJ/cm

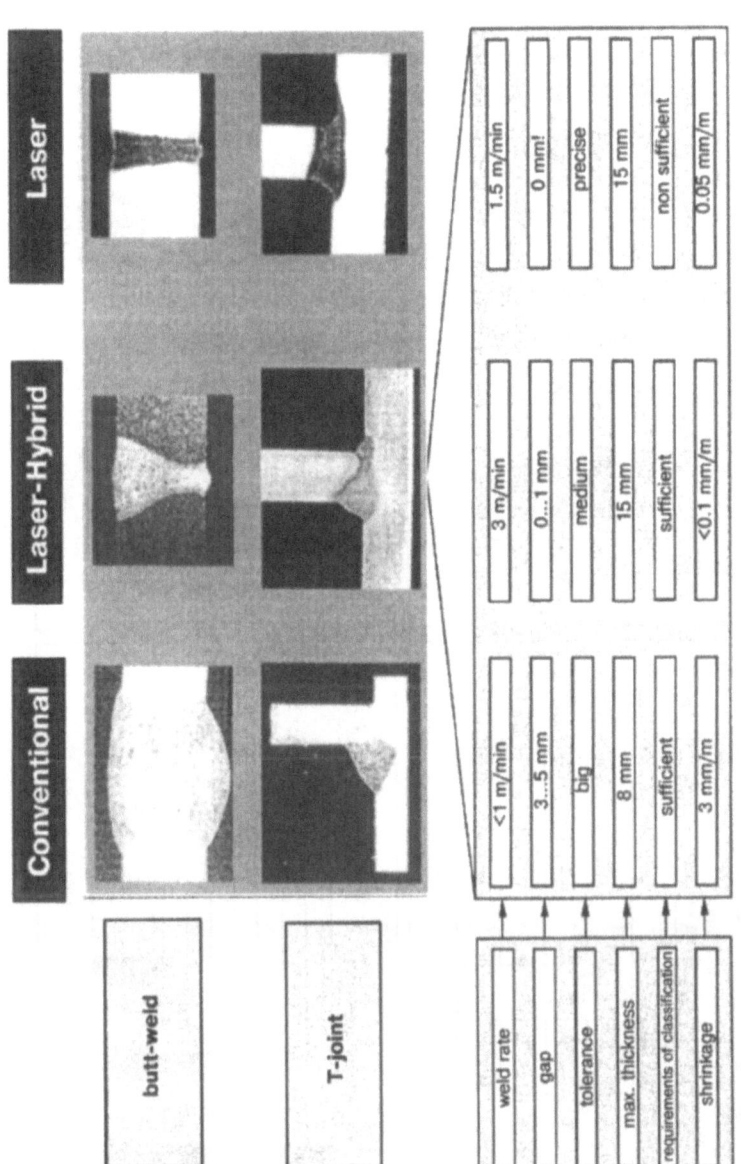

Figure 4.12 Results of various welding processes for butt welds and T-joints.

Table 4.2 Comparison of relative panel shrinkage T-joints, plate 1000× 1000× 5 mm, profile HP 100× 6.

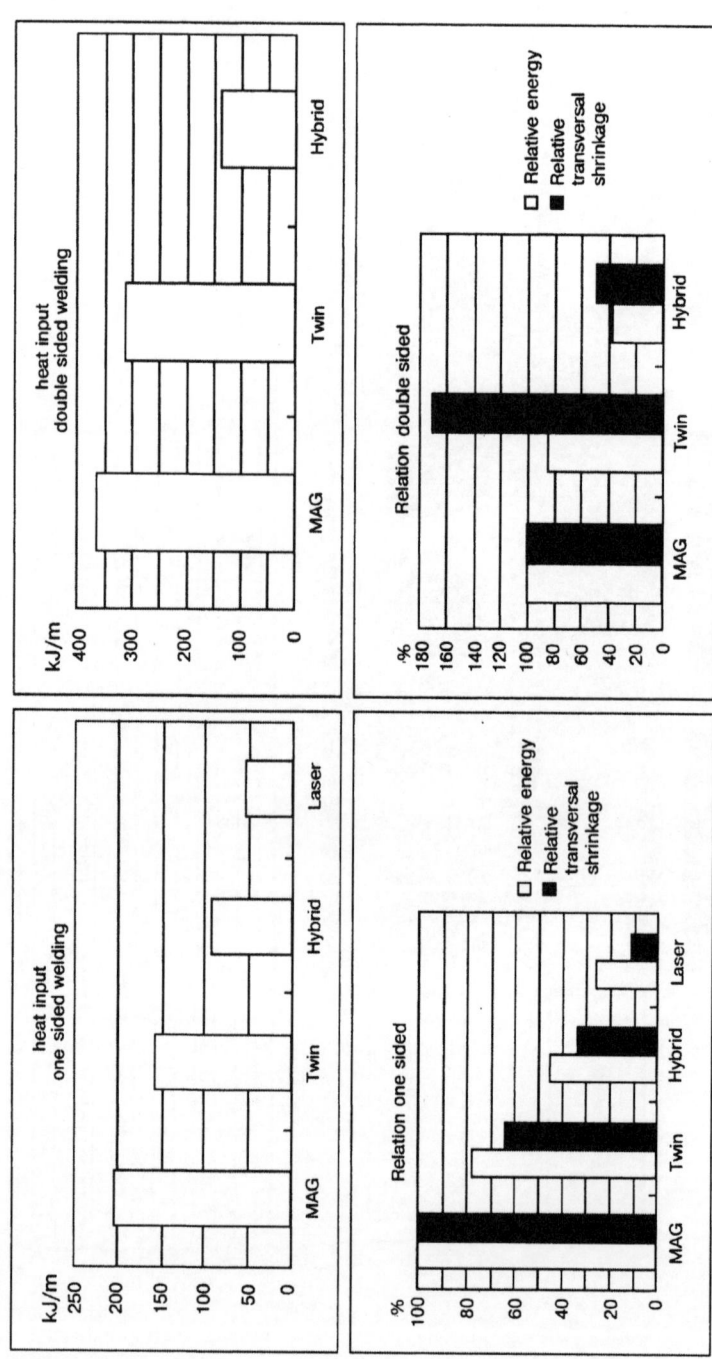

Table 4.3 Welding conditions for laser + GMA welding of A5083 and A6061 aluminium plates [146].

	Shielding box	Torch angle (deg)	Base metal
Experiment I	No	41	A5083
Experiment I	Yes	65	A6061

Filler wire	Laser power (kW)	Welding speed (m/min)	Focus (mm)
A5356	5, 10	1.5, 3.0	± 0
A5356 A4043	5, 10	1.5, 1.0	± 0

Arc voltage (V)	Arc current	Shielding gas (l/min)	
26–27	190–210	40 He	10, 20 Ar
23–24	220–250	40 He	20 Ar

laser + GMA welding of aluminium. Efforts in this respect were made not only in Europe, but in Japan, too [146]. The authors estimated, that:

1 The combination of GMA welding with high power (5 to 10 kW) CO_2-laser beam welding in the flat position bead-on-plate welding of A5083 and A6061 aluminium plates seemed to be a promising method from the viewpoint of stable and smooth bead formation. Depths of penetration were also increased a little. Weld cracks could be prevented by suitable selection of chemical composition and types of filler wires.
2 Oxidation of the weld bead surfaces was reduced by using a shielding shroud around the weld pool area. But the bead appearances were still inferior to those in ordinary GMA welding and further improvement is required.
3 Large blowholes or cavities could be reduced in some welding conditions when A4043 filler wire was selected for welding of A6061 base metal. But in this case the probability of weld cracks was increased.
4 There is still a need for elimination of large blowholes or cavities in butt welding or fillet welding of aluminium plates and evaluation of mechanical properties of the welded joints.

Table 4.3 gives the typical welding conditions used in [146] and Figure 4.13 shows the useful position between the arc torch and the laser head. In conclusion to this item we would like to give some remarkable examples of practical application of the laser-hybrid process (CO_2-laser + MIG arc)* for welding of aluminium alloys [information courtesy of Fronius Schweissmaschinen KG (Austria)]. Thus, Figure 4.14 demonstrates advantages of this process in terms of the weld quality (sufficiently large penetration and good geometrical shape) at a very high welding speed. Figure 4.15 shows a cross-section of the overlap joint with a gap for two dissimilar materials of different thickness. It should be noted here that such a material as AlMg3 pressure cast is very difficult to weld in general, whereas the laser-hybrid process provides a sound weld.

---

* In classification of the hybrid processes suggested in Chapter 1 this process can be referred to as laser + GMA process.

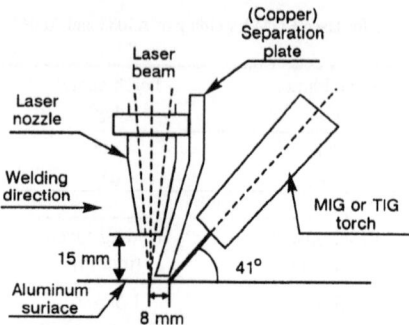

Figure 4.13 Relative position of the laser head and the arc torch [146].

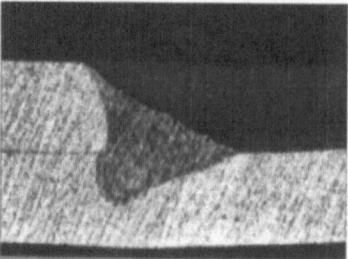

Figure 4.14 Laser-hybrid lap joint of AlMg3 (thickness 2 + 2 mm, welding speed 7 m/min, energy per unit length 800 J/cm) [information courtesy of Fronius Schweissmaschinen KG (Austria)].

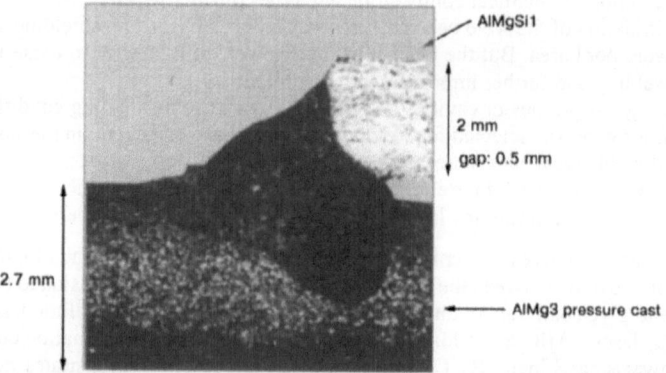

Figure 4.15 Laser-hybrid welding of different materials (filler material: AlSi5, wire diameter 1.2 mm, welding speed 3.6 m/min) [information courtesy of Fronius Schweissmaschinen KG (Austria)].

The laser-hybrid methods have a wide range of technical applications, such as welding of joints of complex geometrical shapes (Figure 4.16) or joints with a large gap between components (Figure 4.17), which are impossible to weld with laser alone. In addition, all advantages of laser welding being preserved, the laser-hybrid processes allow heat input to be dramatically decreased with a simultaneous increase in the welding speed (see, e.g., Table 4.1).

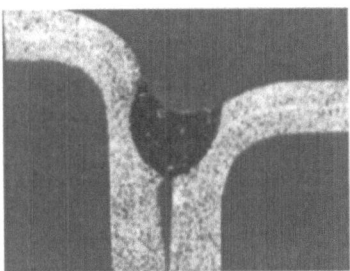

Figure 4.16 Laser-hybrid butt weld between flanges plates (material: AlMg3, plate thickness 1mm, filler material: AlSi5, welding speed 3 m/min, energy per unit length 800 J/m) [information courtesy of Fronius Schweissmaschinen KG (Austria)].

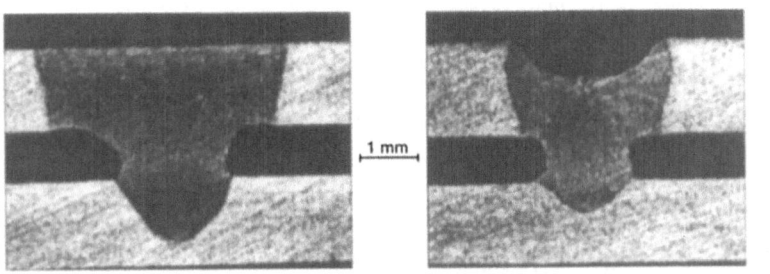

a   b

Figure 4.17 Comparison of laser (a) and laser-hybrid (b) lap joint with the gap (laser beam power 4 kW, MIG arc power 1.2 kW, welding speed 2.4 m/min) [information courtesy of Fronius Schweissmaschinen KG (Austria)].

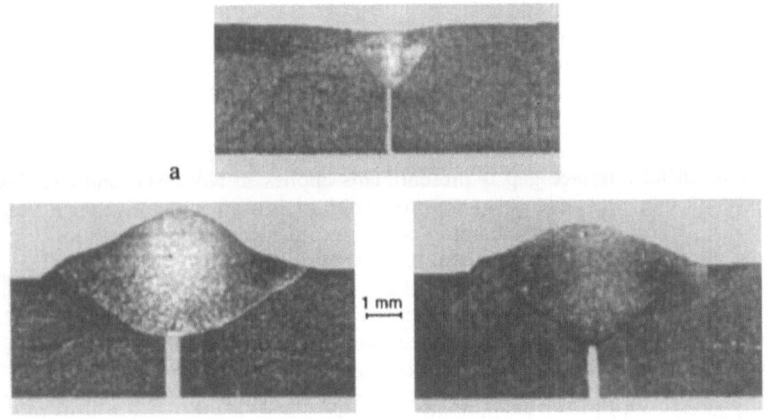

Figure 4.18 Comparison of laser (a), MIG (b) and laser-hybrid (c) butt welds at constant welding speed (1 m/min): a, laser beam power 2.0 kW; b, MIG arc voltage 20 V, arc current 95 A (arc power 1.9 kW), rate of filler wire feeding 11 m/min; c, laser beam power 1.5 kW, rate of filler wire feeding 5.5 m/min (other parameters are the same) [information courtesy of Fronius Sweissmaschinen KG (Austria)].

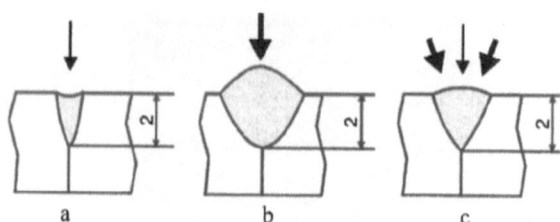

Figure 4.19 Geometry of butt welds (schematically) at constant penetration depth (2 mm) for laser (a), MIG (b) and laser-hybrid (c) welding (parameters are the same as in Figure 4.18) [information courtesy of Fronius Schweissmaschinen KG (Austria)].

Differences between LBW, MIG and laser-hybrid processes, the welding speed being the same, are illustrated in Figures 4.18 and 4.19. Thus, as follows for example from Figure 4.19, at the same welding speed and penetration depth laser-hybrid welding provides a better weld geometry. Moreover, while pure laser beam welding requires a higher accuracy of edge preparation and conventional MIG welding requires a high consumption of filler material, the laser-hybrid process serves as an example of an excellent compromise between the above drawbacks of laser beam welding and MIG welding.

4.4 Laser + Plasma Arc Welding

4.4.1 Feasibility of laser + PA welding

The development of the so-called Plasma Augmented Laser Welding (PALW) technique* has improved CO_2- and Nd:YAG-laser systems. PALW is achieved when a plasma arc is rooted to the point of impingement of the laser beam on the metal surface. Biffin [152] reports on some of the work being carried out and evaluates the capabilities of PALW applied to tailored blanks.

Tests were carried out to determine the laser welding of zinc coated steel in a lap tolerance of the laser beam to poor joint fit configuration has presented a problem due to the low vaporisation temperature of zinc (approximately 900°C) located at the interface between the sheets of material. The application of PALW provides a more stable process operation and reduces the levels of porosity created by the vaporised zinc escaping from the top surface of the weld when no interface gap is present. This applies to Nd:YAG- and CO_2-laser evaluations and is currently undergoing pre-production trials at an automotive manufacturing facility (CO_2-laser). A typical laboratory solution to this problem has been to create a gap between the two sheets of the material using shims. This allows the zinc vapours to exhaust at the interface, but would be impractical as a production solution. Fixturing has been developed to create a gap between the sheets, eliminating the need for shimming. This provides consistent fixturing, producing flat coupon lap welds using the PALW and laser welding techniques, in various thicknesses of coated steel up to 3.5 mm.

Experimentation using a 2 kW Nd:YAG-laser showed that lap welds on 1.0 – 3.5 mm thick coated steels could be produced applying an interface gap of approximately 0.1 – 0.25 mm (conventional laser welding). Top bead shrinkage was found to equate approximately to the volume of material bridging the interface. Gaps less than 0.1 mm resulted in

---

* In classification of the hybrid processes suggested in Chapter 1 this process can be referred to as laser + PA process.

top bead porosity, whereas gaps in excess of 0.25 mm resulted in incomplete fusion between the sheets. PALW trials demonstrated that interface gaps up to 0.5 mm could be tolerated before acceptable fusion was not possible, allowing a more relaxed fit-up regime. Top bead shrinkage was considerably less than the volume of material bridging the joint interface gap. This was a consequence of the thermal action of the plasma arc on the top sheet and was considered to be due to a wider weld zone resulting from plasma augmentation.

4.4.2 Principles, devices and practical examples of laser + PA welding

Using plasma arc, continuous welds were achieved, demonstrating greater process stability and visually reduced defect levels. Both full and partial penetration welds were produced. Full penetration welding speeds were typically between 0.9 and 2.0 m/min at 2 kW laser power with an addition of 1 kW plasma arc energy.

Welding of aluminium alloys

Aluminium alloys are invariably less weldable than automotive steel shields. Two factors contribute to this — the high thermal conductivity of aluminium alloys and their passive oxide layers that have a significantly higher melting temperature than the base alloy and offer poor electrical conductivity. The most successful processes for welding aluminium alloys are those that remove or penetrate the surface oxide layers effectively and provide sufficient rapid heat input to the base material to affect coalescence. The use of aluminium alloys in the automotive industry has been hindered by incompatibility with current production methods. Similarly, laser welding of aluminium alloys is more problematic than for other materials due to the high thermal conductivity and high initial reflectivity to laser radiation. However, with the development of higher power Nd:YAG-lasers, progress is being made to incorporate laser welding of aluminium alloys into auto body application.

The addition of the plasma arc energy (< 1.5 kW) to a 2 kW Nd:YAG-laser was found to enhance welding speed to approximately 100% on aluminium alloys. Material combinations between 1.0 mm and 3.5 mm of 5000 and 6000 series alloys were examined. A pulse output was generally applied to allow effective coupling with aluminium. Weld appearance was comparable to, or improved, compared to the laser alone welds. The addition of the plasma arc required higher levels of shielding gas flow rate. Whereas < 30 l/min of helium was sufficient to provide a clean weld top bead with the arrangement applied for conventional laser welding, increased levels of shielding were essential for PALW. Approximately 45 l/min were necessary to achieve a comparable level of weld cleanness. With the addition of the arc, the laser could be applied in the CW mode to produce a full penetration weld on thinner < 1.5 mm aluminium. This was not possible without the addition of plasma arc energy. Although faster than the laser alone (pulsed operation), welding speed was typically less than applying PALW with the optimum pulsing condition, and varying levels of back reflection occurred resulting in some occurrences of plasma nozzle tip failure [152].

Detailed parameter evaluations have examined the influence on welding performance and quality of the plasma torch orientation, welding direction, plasma arc parameters, shielding gas arrangement, focal characteristics and laser pulse conditions [152]. Pre-production development trials on 3D auto body applications are being carried out.

PALW trials using a 2 kW CO_2-laser with wire filler were carried out for processing aluminium alloy extrusions with consistent but poor fit-up conditions. Butt welds (2.5 mm) were produced at speeds up to 5.0 m/min with an addition of 1.0 kW plasma energy. The addition of the plasma arc also relaxed the wire delivery regime as a consequence of the tolerance provided by the broader arc (as compared to the narrow focus beam). Enhancements to welding speed of 50% were observed and the welding process was more stable than conventional laser welding. A number of systems are currently being installed to industrial applications. Today in operation at Fronius Schweissmaschinen KG (Austria) is a 4.0 kW solid state Haas-Laser with Fronius plasma arc equipment to improve the process.

Plasma arc augmentation contributes to the welding performance of industrial lasers. Benefits are applicable to both CO_2- and Nd:YAG-lasers, welding steels and aluminium alloys. There is considerable economic merit in plasma arc augmentation of welding operations. For a relatively small capital increase, PALW extends the performance of industrial lasers, providing commercial advantages and potential for increased application of lasers in manufacturing. The cost of an additional beam (if available) to achieve similar increases in performance is significant and may not produce other recorded benefits of PALW. An integrated PALW head has been developed and this patented solution is commercially available providing the process robustness demanded from the production environment.

Arrangement for plasma torch and laser beam

Some investigations of combined laser and plasma arc welding were made by the authors of [147] with the following parameters:

laser power Q_1: 10 kW	plasma arc current I: 225 A
welding speed: 1 m/min	sheet thickness: 12 mm
focal distance f: 300 mm	distance torch – laser beam: 100 mm

The plasma torch is arranged in welding direction ahead of the focused laser beam with a forward travel of 100 mm.

A thermo-camera was used to take an instantaneous photograph of the welding process delivering an idea of the temperature field, which is created by the plasma torch. Because the visual angle of the camera to the sheet surface was 30°, the width of the temperature field in the camera picture must be halved. According to the experimental results the temperature is about 150°C on the sheet surface, when the laser beam touches the pre-heated spot.

Influence of welding speed, plasma arc current and laser beam focal distance

The most important welding parameters, which may indicate a significant influence on the pre-heating effect, are the welding speed and the current. The welding speed has also a clear influence on the penetration depth of the laser beam. The welding speed has the most effective influence. In the case of Q_1 = 10 kW and plasma arc current I = 200 A, an increase of the welding speed from 1 m/min up to 2 m/min leads to a decrease of penetration from 13.5 mm down to 10.5 mm. In the current range between 200 A and 275 A no effect of the plasma arc power on the weld dimensions can be determined.

The main criterion of all welding measures is to obtain crack-free welds. At a laser power of 10 kW, a sheet thickness of 12 mm and a welding speed of 1 m/min it was not possible to produce crack-free welds with an optics of f = 300 mm. Applying optics of longer focal distance (f = 380 mm) a crack-free weld could be obtained. Changing the focal distance from 300 to 380 mm has a clear effect on crack formation if the technical measures, such as increase of the focal distance and plasma pre-heating are combined with each other. Each measure alone would not produce crack-free welds as has been proved in tests.

Welding with plasma pre-heating connected with a focal distance $f = 380$ mm can produce completely crack-free radiographs of welded joints (mild steel of 12 mm thickness) at a welding speed of 1 m/min. So far this could not be achieved without pre-heating. The procedure version of laser beam welding with plasma pre-heating supported by other technological measures, such as selection of the beam parameters and the material composition, increases the security of avoiding hot cracking in the centre of the weld.

4.5 Combined Laser Beam + Submerged Arc Welding

4.5.1 Feasibility of laser beam + submerged arc welding

The aim is to use laser systems of medium power for sheet thickness ranges, for which laser power would not be enough for welding the seam, if the residual seam cross-section is welded applying another high power technique, as it is shown in Fig.4.20 and Fig. 4.21. Here it is not necessary to weld in one common working spot with the laser, but the conventional method (submerged arc welding) can be performed in a subsequent process. Because of its high deposit efficiency the submerged arc method is taken into consideration.

4.5.2 Principles, devices and practical examples of laser beam + submerged arc welding

Range of parameters for laser beam welding

Welding tests were carried out using sheets of 12 mm thickness of the steel grade L36TM in order to obtain welding parameters for different types of weld preparation [147]. The sheet thickness of 12 mm has been chosen because it is quite common in shipbuilding industry. On the other hand, this sheet thickness can readily be welded by a laser of medium power (12 kW). Welding penetration tests were done with varying laser power and welding speed and joining welds were executed with milled square-butt weld preparation and milled V-weld preparation with and without filler wire. A high-power laser TLF 12000 has been used. The V-weld was produced as square-edge weld under an enclosed angle of 6°. The following laser parameters are valid for all the results mentioned:

> focus position: $z = 0$ (focus on the surface of the workpiece)
> shielding gas: helium
> quantity of shielding gas: 25 l/min, coaxial supply
> use of filler wire: impingement of the filler wire in the laser beam
> wire diameter: 1.0 mm type SG2.

The laser seam has been welded with and without filler metal. The test material was steel St37 (S235JRG2) with a sheet thickness of 20 mm. The chemical composition of the test material is shown in Table 4.4.

At a sheet thickness of 10 mm both for the web and for the weld opening part, secure fusion and melting of the weld root can be obtained. A concave formation of the upper side of the weld at the sidewalls of the Y-weld opening may be an initiating factor for a crack formation, Figure 4.20. Cracking can be avoided by adding filler metal. The resulting weld

Table 4.4 Chemical composition of steel S235JRG2.

C	Si	Mn	P	S	Al	Mo	Ni	Cr	Co
0.081	0.195	0.587	0.0186	0.0251	<0.0002	0.0276	0.126	0.149	0.013
Cu	Nb	Ti	V	W	As	Zr	Ca	Ce	B
0.469	0.0011	<0.001	<0.0003	<0.008	0.0174	0.0009	0.0004	<0.001	0.0006

shape changes the direction of the dendrite solidification front in a way that no cracking of the concave upper side of the weld would be expected.

Due to the special requirements for accuracy of the weld preparation the only suitable weld shape for the considered example is the V-V-weld. This weld preparation is recommended for all combined laser beam submerged-arc procedures. The lower part of the weld which must be welded by the laser can be adapted with the enclosed angle to the welding method in a way, that by means of the filler wire, the solidification direction can be influenced and cracking can be reduced or even completely avoided. Here the upper part of the weld still is treated with the optimal weld preparation ($30°$ enclosed angle) as has been determined for the conventional method. Tests have shown that an enclosed angle of a laser weld of $7°$ has a favourable influence on the solidification direction and on the formation of crack-free welds. Sporadic cracks at the upper side of the laser weld are again molten by the subsequent submerged-arc welding method and are no longer present after the combined weld has been finished.

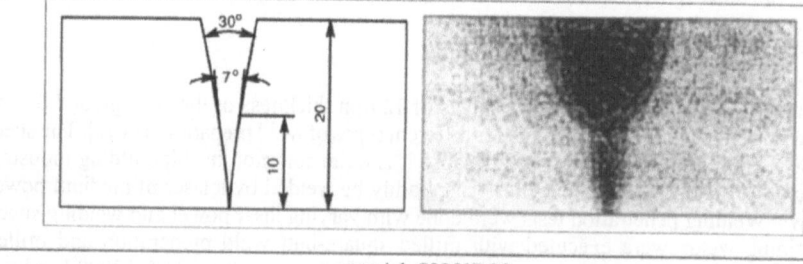

material: S235JRG2	
plate thickness 20 mm	
Laser welding	Submerged arc welding
10.5 kW	38 V/800 A
1 m/min	1 m/min
Filler material: SG2	Filler material: S2
10 m/min	Flux: OP180S

Figure 4.20 Edge preparation and weld cross-section for laser beam + submerged arc welding (root pass: laser beam welded, cover pass: submerged arc welded).

Execution of the combined laser beam + submerged arc welding

The schematic arrangement of a system of combined laser submerged-arc welding is shown in Figure 4.21. It is evident that both laser beam and submerged arc are working on a common moving working unit and have the same welding speed.

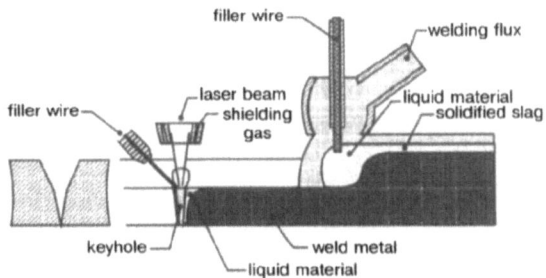

Figure 4.21 Schematic diagram of combined laser beam + submerged arc welding.

At a laser power of 10 kW on the workpiece for a butt weld of a thickness 20 mm at a heat input of 6.3 kJ/cm, according to the welding advisory system WELDWARE, a cooling time of $t_{8/5} = 2.2$ s can be calculated. For this cooling time we get the following mechanical characteristics in the HAZ of the combined laser + submerged arc welding process of steel S235JRG2:

> hardness HV10: 283
> yield strength: 608 N/mm^2
> ultimate tensile strength: 835 N/mm^2
> elongation: 13.8 %
> constriction: 45.6 %.

As a rule, the transformation processes in the laser weld can not be influenced by the subsequent submerged arc welding. For process-technical reasons the distance between the laser beam and the submerged arc is so large that the laser weld has already cooled down below the transformation temperature of the gamma/alpha-transformation and a diffusion-controlled transformation can not occur anymore.

Perhaps a tempering process may be obtained. For this diffusion-controlled process, not only a special temperature is required but also time. The relation between temperature and time is described by the parameter $P = T(\lg t + c)$. As a function of time, action times of more than 20 s are required to create a loss of hardness. If the laser beam + submerged arc welding process is performed in a multi-wire welding, the times required for tempering can be achieved.

The combination of laser beam welding and submerged arc welding offers new possibilities to the user to extend the field of application of given laser beam welding systems for higher ranges of sheet thickness. By a "division" of the weld into a region to be welded by the laser and one region to be welded in the same operation process by a submerged arc, the advantages of both methods can be combined in a way that it is possible to keep the process speed even during welding of thick sheets at such a high level, as if a sheet of 12 mm thickness would be exposed to laser beam welding – that is 1 m/min. In submerged arc multi-wire welding even higher welding speeds can be expected.

A thermal effect on the laser beam welded root of the weld by heat treatment to reduce hardness in the laser weld does not occur. Therefore, the danger of solidification cracking in laser welds can be minimised by metallurgical measures (lower carbon content, low phosphor and sulphur) and possibly by the use of long focal distances in the following optics. Moreover the region of the seam to be welded by laser can be chosen in a way that the danger of cracking should not exist. The residual cross sections can be filled by means of the submerged arc multi- wire technique with adapted weld shape at the same speed.

4.6 Combined Laser Beam + HIGH SPEED Welding

Another version of a combined laser beam arc method, which as with submerged arc method is able to fill large weld volumes with a high melting efficiency, is called the laser beam + HIGH SPEED welding.

In the past years high-power methods have been introduced in industry, which are known as T.I.M.E., Rapid Arc or HIGH SPEED welding (high-power welding with rotating arc). The T.I.M.E. welding is patented and is characterised by the use of a 4-component gas. By means of currents of more than 400 A and wire feed rates of more than 25 m/min the arc is deflected by electro-magnetic forces and pushed into circular motion. The advantages of a rotating arc are:

1 relatively flat and wide penetration profile,
2 common high-current penetration can be avoided,
3 high melting efficiency.

Our experiments have shown that it is not possible to fill the weld cross section without flaws at welding speeds of 1 m/min or more.

The rotating arc causes strong vortices in the weld metal, the melt is "boiling" and transporting gases into the lower region of the weld, where they freeze due to the fast solidification of the molten pool. Flaw-free welds can only be obtained at welding speeds of < 0.70 m/min with sufficient time for degassing.

No subsequent short-time post-heating by the arc occurs, because due to the formation of spatters during T.I.M.E. welding and the high heat radiation, the distance between the laser beam and the arc is so large that the laser weld cools down too quickly.

T.I.M.E. welding in combination with the laser can only be recommended if both methods are executed separately. Root fusion by means of laser beam welding without backing for the molten pool and subsequent welding by the high power method HIGH SPEED can increase the efficiency of the production very much.

4.7 Hybrid CO_2 + High Power Diode Laser Welding

Galvanised car body sheets are highly corrosion-resistant and are thus more and more used in the motor industry. In particular, laser beam welding of overlapping seams causes problems because of the two interior zinc layers and has not found a satisfactory solution. The boiling point of zinc is approximately 900°C. During the welding process temperatures in the range of 1500–3000°C cause evaporation of zinc.

This leads to fusion ejections and pores in the welding seam, since the degassing is made more difficult by the close steam capillary. Such welding seams possess reduced strength and a bad surface quality, which results inl intensified corrosion.

This problem can be solved by expanding the vapour capillary occurring in laser beam welding, accompanied by a longer existence of the weld pool due to the action of an additional energy source. A high power diode laser (HPDL) is very useful because of its good local heat input without producing plasma that weakens the CO_2-laser radiation and because it can simply be integrated into the given laser welding systems.

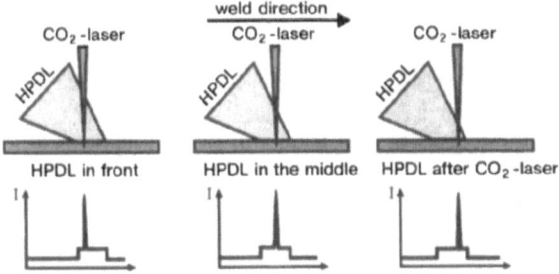

Figure 4.22 Relative position between CO_2-laser and high power diode laser beams.

The overlay of the focused CO_2-laser by a rectangular intensity distribution of several square millimetres of the high power diode laser enables an expansion of the steam capillary, to extend the duration of the weld pool and thus to accelerate degassing of the zinc steam. The temporal frame is extended in a way such that this has finished before the solidification of the melt. The process rate mainly given by the CO_2-laser can be slightly increased.

Different fundamental investigations were carried out on the combination of both energy sources. Laser power and related positions were changed. Figure 4.22 shows schematically the relative position between the CO_2-laser and HPDL beams. Figure 4.23 shows the result of these various relative positions for hybrid welding of zinc coated car body sheets. The best result is obtained in the case where the diode laser beam is running after the CO_2-laser beam.

It turned out that the welding depth does not substantially improve by additional energy at the same feed rate. It is however obvious that the welding seam is expanding to the top. A forward-moving diode laser beam has a larger depth effect, while a following diode laser beam better reduces undercuts, Figure 4.23. For a specific case of application there is the possibility to implement an optimal arrangement. For the tested process parameters and plate thickness of 0.75 mm to 1.25 mm the formation of pores and ejections could be avoided in all cases.

Also the next examples show the positive influence of the hybrid CO_2 + high power diode laser process on the cross-section of weldments in the case of welding steel, Figure 4.24 and AlMgSi1, Figure 4.25.

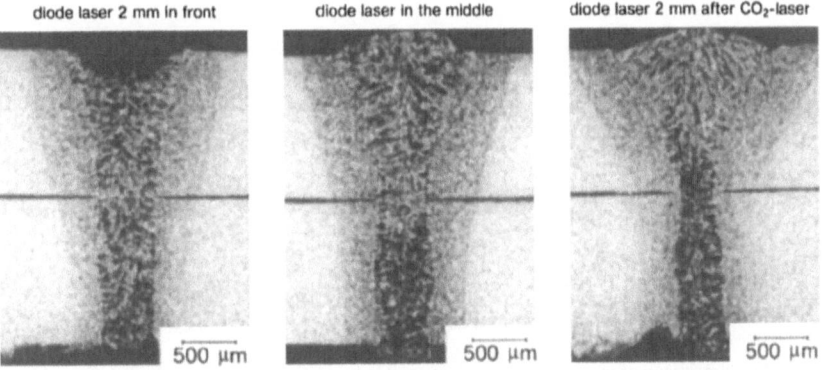

Figure 4.23 Hybrid CO_2-laser (3.5 kW) + high power diode laser (1.5 kW) welding of zinc coated car body sheets.

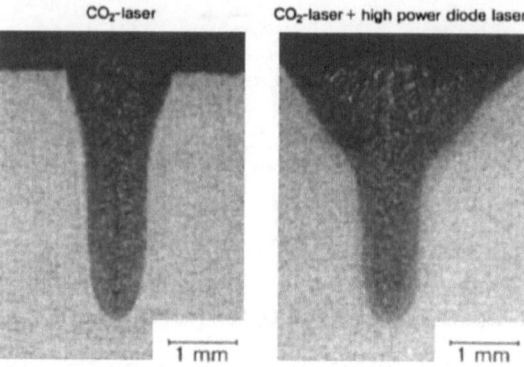

Figure 4.24 Comparison of CO_2-laser welding and hybrid CO_2 + high power diode laser welding of steel TStE355 (CO_2-laser power 2.0 kW, HPDL power 1.5 kW, welding speed 2.5 m/min).

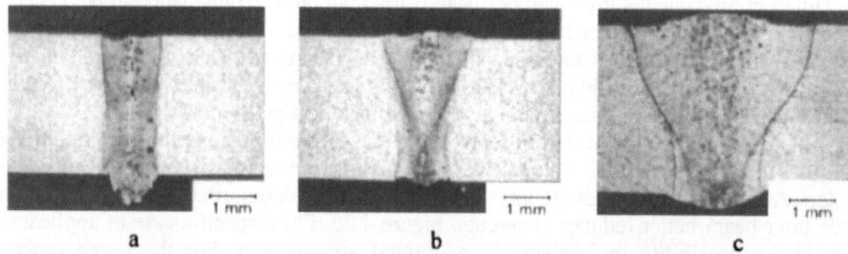

Figure 4.25 Comparison of CO_2-laser (a), CO_2 + high power diode laser (b) and CO_2- laser + plasma arc welding of 3 mm thick AlMgSi1 sheets: a, CO_2-laser power 3.5 kW, welding speed 5 m/min, energy per unit length 42 kJ/m; b, CO_2-laser power 3.5 kW, HPDL power 1.5 kW, welding speed 5 m/min, energy per unit length > 60 kJ/m; c, CO2- laser power 2.4 kW, plasma arc current 50 A, welding speed 3 m/min, energy per unit length > 68 kJ/m.

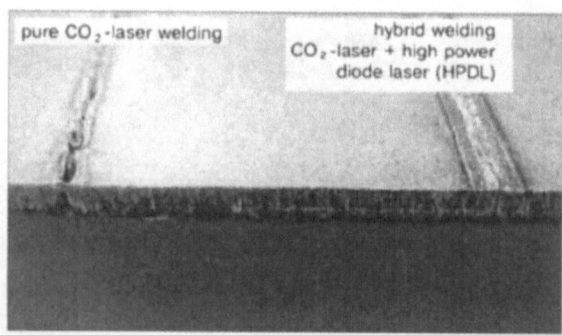

Figure 4.26 Comparison of pure CO_2-laser welding and hybrid CO_2 + high power diode laser welding of zinc coated car body sheets (CO_2-laser power 2.0 kW, HPDL power 1.5 kW, welding speed 4 m/min, sheet thickness 0.75 mm) [153].

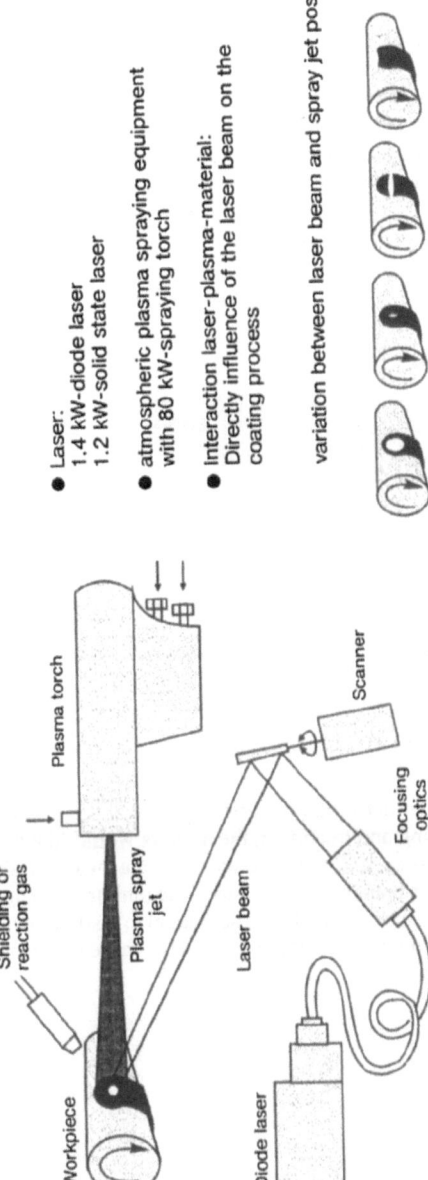

Figure 4.27 Laser aided thermal spraying: technique and principle.

The positive influence of the solid state laser for smoothing the weld surface is remarkable. A view is given on the surface of car body sheet, zinc coated, welded with only CO_2-laser and also with the hybrid method, Figure 4.26 [153].

4.8 Laser Aided Thermal Spraying and Powder Cladding

There are more useful possibilities for using laser as an additional process variable. One method already being operated is a laser aided thermal spraying. The two figures, Figure 4.27 and Figure 4.28, show very impressive details of this spraying process. This is another kind of hybrid CO_2-laser and plasma process*. There are a lot of techniques for various materials to improve the surfaces and to get new surface characteristics. At the same time the productivity is rising to a certain extent.

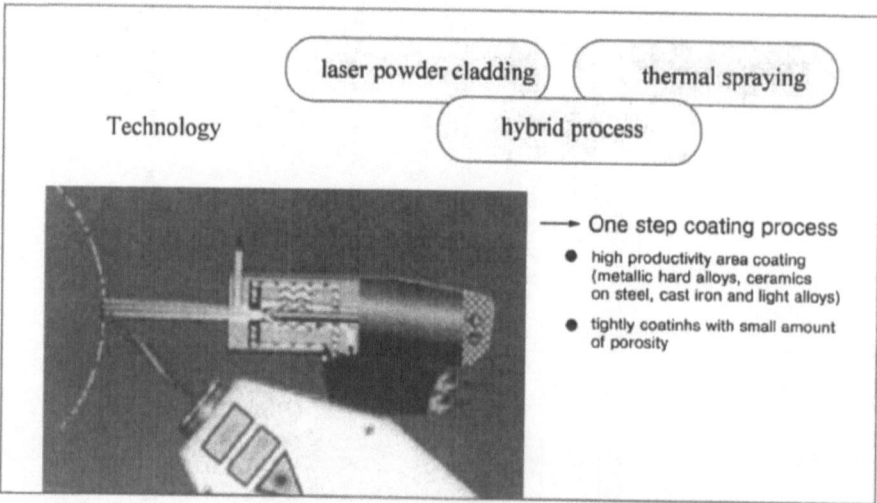

Figure 4.28 Laser aided thermal spraying: technology.

Completing this chapter dedicated to various practical applications of hybrid laser-arc processes of material joining and treatment, the authors would like to express the hope that in the near future such processes will take an adequate place in ship building, aircraft, automotive and many other branches of industry. The authors also hope that this book will give a new impetus both to experimental and theoretical investigations of laser-arc processes and, eventually, will promote the development of new, more efficient hybrid technologies.

---

* In classification of the hybrid processes suggested in Chapter 1 this process can be referred to as laser + PJ process.

References

1. Steen, W. M. *Methods and apparatus for cutting, welding, drilling and surface treating.* Pat. No.1547172, Great Britain, Int. Cl. B23K 26/00, 9/00, published 06.06.79.

2. Steen, W. M. *Improvements in methods and apparatus for cutting, welding and surface treating.* Pat. No.1600796 (Addition to Pat. No.1547172), Great Britain, Int. Cl. B23K 26/00, 9/00, published 21.10.81.

3. Steen, W. M. *Methods and apparatus for cutting and welding.* Pat. No.4167662, USA, Int. Cl. B23K 9/00, published 11.09.79.

4. Steen, W. M. *Verfahren und vorrichtung zur bearbeitung von werkstucken mittels eines laserstrahls.* Pat. No.2813642, Germany, Int. Cl. B23K 26/00, published 04.10.79.

5. Eboo, M., Steen, W. M. and Clarke, J. (1978) Arc augmented laser welding, In: *Advances in welding processes*, Proceedings of the 4th Int. Conf., England, Harrogate, vol.1, pp. 257–265.

6. Steen, W. M. and Eboo, M. (1979) Arc augmented laser welding, *Metal Construction*, **11**, No. 7, pp. 332–335.

7. Clarke, J. and Steen, W. M. (1979) Arc augmented laser cutting, In: *Proceedings of the Laser 1979 Conf.*, Germany, Munich, p. 247.

8. Steen, W. M. (1980) Arc augmented laser processing of materials, *J. of Appl. Phys.*, **51**, No. 11, pp. 5636–5641.

9. Alexander, J. and Steen, W. M. (1980) Penetration studies on arc augmented laser welding, In: *Proceedings of Int. Conf. on Welding Research in the 1980s*, Japan, Osaka, pp. 121–129.

10. Alexander, J. and Steen, W. M. (1981) Arc augmented laser welding process - variables, structure and properties, In: *Join. Metals. Pract. and Perform.*, Proceedings of Spring Resident. Conf. No.18, England, Warwick, vol.1, pp. 155–160.

11. Mazumder, J. and W.M.Steen, W.M. (1981) Laser welding of steels in can making, *Welding J.*, **60**, No.6, pp. 19–25.

12. Hamasaki, M. *Welding method combining laser welding and TIG welding.* Claim 56-49195, Japan, Int. Cl. B23K 26/00, 9/00, published 20.11.81.

13. Shato, A. and Kavaguti, S. *Combined welding method.* Claim 58-184081, Japan, Int. Cl. B23K 26/00, published 27.10.83.

14. Nakamura, T., Kono, R. Minamida, K. and Takanatsu, H. *Laser welding method.* Claim 58-184082, Japan, Int. Cl. B23K 26/12, published 27.10.83.

15. Nakamura, T., Kono, R. and Fusita, H. *Laser welding method.* Claim 58-184084, Japan, Int. Cl. B23K 26/12, published 27.10.83.

16. Nakamura, T. and Kono, R. *Laser welding method.* Claim 58-184085, Japan, Int. Cl. B23K 26/12, published 27.10.83.

17. Minamida, K., Takafusi, H., Nakamura, T. and Fusita, H. *Welding method with high concentration of energy.* Claim 59-42196, Japan, Int. Cl. B23K 26/12, published 08.03.84.

18. Shugiyama, S., Nakayama, K. and Shano, N. *Method of welding.* Claim 59-232690, Japan, Int. Cl. B23K 31/06, 9/02, published 27.12.84.

19. Hamasaki, M. *Welding method combining laser welding and MIG welding.* Claim 59-66991, Japan, Int. Cl. B23K 26/12, 9/16, published 16.04.84.

20. Hamasaki, M. *Welding method combining laser welding and MIG welding.* Claim 60-8916, Japan, Int. Cl. B23K 26/12, 9/16, published 16.04.84.

21. Hamasaki, M. *Welding method combining laser welding and MIG welding.* Pat. No.4507540, USA, Int. Cl. B23K 27/00, published 26.03.85. Claims priority, application Japan, 06.10.82, No.57-175883.

22. Hashiura, M., Uno, T. and Shusuki, S. *Laser working device.* Claim 60-106688, Japan, Int. Cl. B23K 26/12, 26/14, published 12.06.85.

23. Kanehara, Y. *Working head in working device using laser light.* Claim 60-154894, Japan, Int. Cl. B23K 26/14, published 14.08.85.

24. Hoshinouchi, S., Kanaoka, M. and Fukada, A. *Laser working apparatus.* Claim 60-216989, Japan, Int. Cl. B23K 26/14, published 30.10.85.

25. Hoshinouchi, S., Kanaoka, M. and Fukada, A. *Laser-beam operated machining apparatus.* Pat. No.4689466, USA, Int. Cl. B23K 26/00, published 25.08.87. Claims priority, application Japan, 10.04.84, No.59-71295.

26. Kanaoka, M. and Hoshinouchi, S. *Laser beam working apparatus.* Claim 60-234782, Japan, Int. Cl. B23K 26/00, 26/14, published 21.11.85.

27. Ono, M. *Laser welding method.* Claim 61-232079, Japan, Int. Cl. B23K 26/00, 9/00, 9/16, published 16.10.86.

28. Omay, T. *Laser welding method.* Claim 63-30193, Japan, Int. Cl. B23K 26/00, 9/16, published 08.02.88.

29. Khalboshin, A. P., Kurochkin, Yu. V., Lubchenko, A. M. and Zverev, A. A. (1990) Increase in speed of laser cutting of metals by plasma heating, *Svar. Proizvod.*, No. 12, pp. 3–5.

30. Walduck, R. P. and Biffin, J. (1994) Plasma arc augmented laser welding, *Welding and Metal Fabrication*, **62**, No. 4, pp. 172–176.

31. Cullison, A. (1994) Two processes together are proving better than one, *Welding J.*, **73**, No. 11, p. 16.

32. Tusek, J. (1996) Sinergic operation of welding arc and laser beam-for practical application or for scientific research only?, *Varilna tehnika*, Ljubljana, **45**, No. 2, pp. 39–46.

33. Yoneda, M. and M.Katsumura, M. (1989) Laser hybrid processing, *J. Jap. Weld. Soc.*, **58**, No. 6, pp. 427–434.

34. Merard, R. and Luciani, P. Y. (1983) Sodure par faisceau laser a CO2 de faible puissance, associe a une sourse auxiliaire de chauffage, In: *Proceedings of 3ᵉᵐᵉ Colloq. Int. Soudage et Fusion Faisceau Electrons et Laser, France*, Lyon, pp. 561–568.

35. Luciani, P. Y., Charissoux, C. and Calret, J. N. (1986) CO_2-laser auxiliary source coupling: application to welding, In: *Proceedings of the 3rd Int. Conf. Lasers Manuf.* [LIM-3], France, Paris, pp. 117–123.

36. Hamasaki, M. (1983) Method of material processing by using laser, *Weld. Technique*, **31**, No. 12, pp. 64–69.

37. Hamasaki, M. (1984) New method of laser welding — laser welding with using MIG welding, *Pip. Eng.*, **26**, No. 2, pp. 51–55.

38. Hamasaki, M. and Katsumura, M. (1984) Welding method combining laser radiation and electric arc, *Tool Eng.*, **28**, No. 4, pp. 25–29.

39. Matsuda, J., Utsumi, A., Katsumura, M., Hamasaki, M. and Nagata, S. (1988) TIG or MIG arc augmented laser welding of thick mild steel plate, *Joining and Materials*, **1**, No. 1, pp. 31–34.

40. Matsuda, J. (1989) Laser-MIG welding of thick mild steel plates, *J. High. Temp. Soc.*, **14**, No. 5, pp. 234–239.

41. Diebold, T. P. and Albright, C. E. (1984) "Laser-GTA" welding of aluminum alloy 5052, *Welding J.*, **63**, No. 6, pp.18–24.

42. Devletian, J. H. (1987) *Arc-augmented laser welding of aluminium*, Welding Research Council, New-York [Final Rept].

43. Wendelstorf, J., Decker, I. and Wohlfahrt, H. (1994) Laser-enhanced gas tungsten arc welding (laser-TIG), *Weld. World*, **34**, pp.395–396.

44. Nakata, K., Kurosawa, T. and Yoshikawa, M. (1996) Sumitomo jukikai giho, *Techn. Rev.*, **44**, No. 131, pp 25–28.

45. Moeniralam, Z. and Luijendijk, T. (1996) Wisselwerking tussen laserlassen en boog-lassen, *Lastechniek*, **62**, No. 7–8, pp. 3–6.

46. Irving, B. (1994) Automotive engineers plunge into tomorrow's joining problems, *Welding J.*, **73**, No. 11, pp. 47–50.

47. Uglov, A. A. (1985) 1st All-Union Conf. on Laser Metallurgy and Laser-Plasma Processing (chronicle), *Fizika i khimija obrabotki materialov*, No. 5, p. 143.

48. Panchev, V. A., Plekin, V., A., Gening, P. E., et al (1990) Industrial application of lasers in pipe welding fabrication, *Svar. Proizvod.*, No.12, pp. 2–3.

49. Gutman, M. B., Rubin, G. K. and Seleznev, Yu. N. (1986) Laser- plasma-arc working of metal workpieces, *Avtomobil'naya Promishlenost'*, No.10, pp. 32–33.

50. Seleznev, Yu. N. and Zhuravel, V. M. (1988) Laser-arc working of workpieces, *Avtomobil'naya Promishlenost'*, No.2, p. 23.

51. Borodachev, A. S., Zhuravel, V, M. and Yu.N.Seleznev, Yu. N. (1988) Analysis of laser- arc metal working processes, *Electrotekhnika*, No.11, pp. 16–18.

52. Seleznev, Yu. N. (1988) Application of absorptive coatings at laser-arc heat treatment, *Electrotekhnika*, No.11, pp. 18–20.

53. Duley, W. W. (1983) *Laser processing and analysis of materials*, New York, Plenum Press, 504 pp.

54. Gorny, S. G., Lopota, V. A., Redozubov, V. D. and Smirnov, V. S. (1987) Characteristics of DC arcs of straight polarity at a simultaneous heating of metal with laser beam, *Avtomat. Svarka*, No.11, pp. 73–74.

55. Bashenko, V. V., Gorny, S. G., Lopota, V. A. and Redozubov, V. D. (1988) *Laser-arc welding of metals*, Leningrad, Leningrad House of Scientific and Technical Propaganda, 22 pp.

56. Gorny, S. G., Lopota, V.A. and Redozubov, V. D. (1989) Research of electric characteristics of arc in laser-arc welding, *Svar. Proizvod.*, No. 1, pp. 28–29.

57. Gorny, S. G., Lopota, V.A., Redozubov, V. D. and Smirnov, V. S. (1989) Peculiarities of metal heating in laser-arc welding, *Avtomat. Svarka*, No. 1, pp. 73–74.

58. Petrov, A. L., Zaikin, A. E., Kanavin, A. P., et al (1988) Physical fundamentals of a laser- arc effect on metals. In: *Proceedings of 4eme Colloq. Int. Soudage et Fusion Faisceau Electrons et Laser*, France, Cannes, pp. 203–210.

59. Gureev, D. M., Zaikin, A. E., Zolotarevsky, A. B., *et al.* (1989) Method of laser-arc material processing and its application, In: *Transactions of the Physical Institute of the USSR Academy of Science*, Moscow, Nauka, **198**, pp. 41–61.

60. Bibik, O. B., Brodyagin, V. N. and Pokladov, Yu. P. (1990) Peculiarities of interaction of laser irradiation and welding electric arc in application to combined laser-arc welding process, *Fizika i khimija obrabotki materialov*, No. 2, pp. 95–98.

61. Zhao, J. R., Zhang, S. B., D.Sun, D., *et al.* (1990) Research of a new welding technique - double heat source laser-arc, *IIW Doc.*, No. XII-1187-90, pp. 375–390.

62. Laser teams with tungsten arc to speed sheetmetal welding, (1991) *Weld. Design and Fabr.*, **64**, No. 8, p. 14.

63. Shengsun, H., Shaobin, Z., Dengping L., et al (1991) A study of arc augmented laser welding, *IIW Doc.*, No. XII-1267-91, pp. 207–213.

64. Cui, H. (1991) *Untersuchung der wechselwirkungen zwitchen schweisslichtbogen und fokussiertem laserstrahl und der anwendungsmoglichkeiten kombinierter laser-lichtbogen technik*, GKSS [Rept.], No. E62, pp. 1–12.

65. Gvozdetsky, V. S., Krivtsun, I. V., Svirgun, A. A. and Chizhenko, M. I. (1989) *Constricted laser-arc discharge*, Kiev, E.O.Paton Electric Welding Institute, 30 pp.

66. Gvozdetsky, V. S., Krivtsun, I. V., Svirgun, A. A. and Chizhenko, M. I. (1990) Estimation of laser radiation influence on characteristics of the arc column plasma inside plasma torch channel, *Avtomat. Svarka*, No. 8, pp. 8–14.

67. Dowden, J. M., Kapadia, P. D. and Sibold, D. (1991) Mathematical modelling of laser welding, *Int. J. Joining of Materials*, **3**, pp. 73–78.

68. Gvozdetsky, V. S., Krivtsun, I. V. and Chizhenko, M. I. (1991) *Calculated investigation of discharge characteristics inside channel of the laser-arc plasma torch*, Kiev, E.O.Paton Electric Welding Institute, 42 pp.

69. Gvozdetsky, V. S., Krivtsun, I. V. and Chizhenko, M. I. (1991) Laser beam and electric arc plasma interaction, In: *Proceedings of the 8th All-Union Conf. on Low-Temperature Plasma*, Minsk, pp. 31–32.

70. Gvozdetsky, V. S., Krivtsun, I. V. and Chizhenko, M. I. (1992) Numerical investigation of the laser-arc discharge characteristics in argon flow, In: *Proceedings of the 1st Int. Conf. on Arc Physics*, Kiev, pp. 24–25.

71. Gvozdetsky, V. S., Krivtsun, I. V. Chizhenko, M. I. and Yarinich, L. M. (1995) Laser-arc discharge: Theory and application, *Welding and Surfacing Rev.*, **3**, Harwood, 148 pp.

72. Zhukov, M. F., Smolyakov, V. Ya. and Urukov, B. A. (1973) *Electric arc gas heaters (plasma torches)*, Moscow, Nauka, 232 pp.

73. Zhukov, M. F., Urukov, B. A., Engelsht, V. S., *et al.* (1987) *The theory of thermal electric arc plasma.* Part 1. Mathematical methods of plasma investigation, Novosibirsk, Nauka, 288 pp.

74. Zhukov, M. F., Urukov, B. A., Engelsht, V. S., *et al.* (1987) *The theory of thermal electric arc plasma.* Part 2. Non-stationary processes and radiation heat transfer in thermal plasma, Novosibirsk, Nauka, 288 pp.

75. Engelsht, V. S., Gurovich, V. Z., Desyatkov, G. A., *et al.* (1990) *The theory of electric arc column*, Novosibirsk, Nauka, 376 pp.

76. Raizer, Yu. P. (1977) *Laser induced discharge phenomena*, New York, Plenum Press, 308 pp.

77. Raizer, Yu. P. (1980) Optical discharges, *Uspekhi Fizich. Nauk*, **132**, pp. 549–581.

78. Gerasimenko, M. V., Kozlov, G. I. and Kuznetsov, V. A. (1983) Laser plasma torch, *Kvant. Elektronika*, **10**, No. 4, pp. 709–717.

79. Katsenelenbaum, B. Z. (1966) *High-frequency electrodynamics*, Moscow, Nauka, 240 pp.

80. Yariv, A. (1975) *Quantum electronics*, New York, Wiley, 488 pp.

81. Landau, L. D. and Lifshits, E. M. (1982) *Electrodynamics of continuous media*, Moscow, Nauka, 620 pp.

82. Paskonov, V. M., Polezhaev, V. I. and Chudov, L. A. (1984) *Numerical modelling of heat and mass transfer*, Moscow, Nauka, 286 pp.

83. Samarsky, A. A. and Nikolaev, E. S. (1978) *Solution of grid equations by numerical methods*, Moscow, Nauka, 591 pp.

84. Gvozdetsky, V. S., Krivtsun, I. V. and Chizhenko, M. I. (1988) Calculation of thermal properties and transport coefficients of Ar-He plasma, In: *Application of mathematical methods in welding*, Kiev, E.O.Paton Electric Welding Institute, pp. 21–28.

85. Godnev, I. N. (1956) *Calculation of thermodynamic functions using molecular data*, Moscow, OGIZ, 412 pp.

86. Kasabov, G. A. and Eliseev, V. V. (1973) *Spectroscopic tables for low-temperature plasma*. A handbook, Moscow, Atomizdat, 160 pp.

87. Lick, W. J. and Emmons, H. W. (1965) *Transport properties of helium from 200 to 50 000 K*, Cambridge, Massachusetts: Harvard Univ. Press, 106 pp.

88. Drellischak, K. S. (1964) *Partition functions and thermodynamic properties of high temperature gases*, AEDC, TDR-64-22, 10, pp. 1–148.

89. Katsnelson, S. S. and Kovalskaya, G. A. (1985) *Thermodynamic and optical properties of argon plasma*, Novosibirsk, Nauka, 148 pp.

90. Lancaster, J. F. (1984) *The physics of welding*, Oxford, Pergamon Press, 297 pp.

91. Zhdanov, V. M. (1982) *Transport phenomena in multicomponent plasma*, Moscow, Energoizdat, 176 pp.

92. Hirschfelder, J. O., Curtiss, Ch. F. and Bird, R. B. (1954) *Molecular theory of gases and liquids, New York*, Jonh Wiley and Sons Inc., 930 pp.

93. Krinberg, I. A. (1965) Effect of ionisation on heat conductivity of plasma, *Teplofizika Vysokikh Temperatur*, **3**, No. 6, pp. 838–844.

94. Chapman, S. and Cowling, T. G. (1939) *The mathematical theory of non-uniform gases*, Cambridge, 510 pp.

95. Fon, W. C., Berrington, K. A. and Hibbert, A. (1981) Elastic scattering of electrons from inert gases: I. Helium, *J. Phys. B.: At. Mol. Phys.*, **14**, No. 2, pp. 205–219.

96. Fon, W. C., Berrington, K. A., Burke, P. G. and Hibbert, A. (1983) Elastic scattering of electrons from inert gases: III. Argon, *J. Phys. B.: At. Mol. Phys.*, **16**, No. 2, pp. 307–321

97. Smirnov, B. M. (1968) *Atomic collisions and unit processes in plasma*, Moscow, Energoatomizdat, 363 pp.

98. Radtsig, A. A. and Smirnov, B. M. (1986) *Parameters of atoms and atomic ions*. A handbook, Moscow, Energoatomizdat, 344 pp.

99. Vargaftik, N. B. (1978) *Thermal properties of gases and liquids*. A handbook, Moscow, Atomizdat, 730 pp.

100. Devoto, R. S. (1973) Transport coefficients of ionized argon, *Phys. Fluids*, **16**, No. 5, pp. 616–623.

101. Devoto, R. S. and Li, C. P. (1968) Transport coefficients of partially ionized helium, *J. Plasma Phys.*, **2**, No. 1, pp. 17–32.

102. Ginzburg, V. L. (1967) *Propagation of electromagnetic waves in plasma*, Moscow, Nauka, 684 pp.

103. Hora, H. (1981) *Physics of laser driven plasmas*, New York, Wiley, 272 pp.

104. Biberman, L. M. and Norman, H. E. (1967) Continuous spectra of atomic gases and plasma, *Uspekhi Fizicheskikh Nauk*, **91**, No. 2, pp. 193–246.

105. Raizer, Yu. P. (1987) *Physics of gas discharge*, Moscow, Nauka, 592 pp.

106. Batenin, V. M. and Minaev, P. V. (1977) About radiation of dense low-temperature plasma of inert gases, *Teplofizika Vysokikh Temperatur*, **15**, No. 3, pp. 647–649.

107. Batenin, V. M. and Minaev, P. V. (1969) Continuous radiation of low-temperature argon plasma, *Teplofizika Vysokikh Temperatur*, **7**, No. 4, pp. 604–609.

108. Kozlov, G. I., Kuznetsov, V. A. and Masyukov, V. A. (1974) Radiative losses from an argon plasma and a radiation model of a continuous optical discharge, *Zurnal Experimentalnoi i Tekhnicheskoi Fiziki*, **66**, No. 3, pp. 954–964.

109. Evans, D. L. and Tankin, R. S. (1967) Measurement of emission and absorption of radiation by an argon plasma, Phys. Fluids, 10, No. 6, pp. 1137–1144.

110. Born, M. and Wolf, E. (1968) *Principles of optics*, Oxford, Pergamon Press, 719 pp.

111. Vedenov, A. A. and Gladush, G. G. (1985) *Physical processes at laser processing of materials*, Moscow, Energoatomizdat, 208 pp.

112. Glickstein, S. S. (1979) Arc modelling for welding analysis, In: *Proceedings of Int. Conf. on Arc Phys. and Weld Pool Behav.*, London, Paper 5, 16 pp.

113. Walduck, R. P. *Enhanced laser beam welding.* Pat. No.5866870, USA, Int. Cl. B23K 10/00, 26/00, published 02.02.99.

114. Paton, B. E. (1995) Upgrading of welding methods - one of the ways to improve quality and cost effectiveness of welded joints, *Avtomat. Svarka*, No. 11, pp. 3–11.

115. Krivtsun, I. V. and Chizhenko, M. I. (1997) Fundamentals of calculating laser-arc plasma torches, *The Paton Welding Journal*, **9**, No. 1, pp. 19–24.

116. Dykhno, I. S., Krivtsun, I. V. and Ignatchenko G. N. *Combined laser and plasma arc welding torch.* Pat. No.5700989, USA, Int. Cl. B23K 26/00, 10/00, published 23.12.97.

117. Dykhno, I., Parneta, I., Ignatchenko, G. and Chizhenko, M. *Combined laser and plasma arc welding torch.* Pat. No.5705785, USA, Int. Cl. B23K 26/00, published 06.01.98.

118. Bushma, A. I. and Krivtsun, I. V. (1992) Peculiarities of fine-dispersion ceramic particles heating by laser irradiation, *Fizika i khimija obrabotki materialov*, No. 2, pp. 40–48.

119. Zhukov, M. F., Kozlov, N. P., Pustogarov, A. V., *et al.* (1982) *Near-electrode processes in arc discharges*, Novosibirsk, Nauka, 157 pp.

120. Dyuzhev, G. A., Nemchinsky, V. A., Shkolnik, S. M. and Yuriev, V. G. (1983) Anode processes in high-current arc discharge, *Khimija plasmy*, **10**, pp. 169–209.

121. Gvozdetsky, V. S., Korchinsky, G. M., Krivtsun, I. V., *et al.* (1986) Calculation of power factors of absorption and reflection of electromagnetic radiation in laser welding, *Avtomat. Svarka*, No. 5, pp. 33–37.

122. Gvozdetsky, V. S., Korchinsky, G. M., Krivtsun, I. V., *et al.* (1987) To calculation of absorption and reflection factors of the opical beam in laser treatment of materials, *Avtomat. Svarka*, No. 1, pp. 70–71.

123. Anderson, D. A., Tannehill, J. C. and Pletcher, R. H. (1984) *Computational fluid mechanics and heat transfer*, New York, Hemisphere Publishing Corporation, 726 pp.

124. Favre, A. (1965) Equations des gaz turbulents compressibles: I. Formes generales, *J. Mecanique*, **4**, pp. 361–390.

125. Loitsyansky, L. G. (1973) *Mechanics of liquid and gas*, Moscow, Nauka, 847 pp.

126. Prandtl, L. (1926) Ueber die ausgebildete turbulenz, In: *Proceedings of the 2nd Int. Congress for Applied Mechanics*, Zurich, pp. 62–74.

127. Deissler, R. G. (1959) *Analysis of turbulent heat transfer, mass transfer and friction in smooth tubes at high Prandtl and Schmidt numbers*, NASA Report, 210 pp.

128. Madni, I. K. and Pletcher, R. H. (1975) Prediction of turbulent jets in coflowing and quiescent ambients, *J. Fluids Eng.*, **97**, pp. 558–567.

129. Ievlev, V. M. (1975) *Turbulent motion of high-temperature continuous media*, Moscow, Nauka, 254 pp.

130. Biberman, L. M., Vorobyov, V. S. and Yakubov, I. T. (1969) Non-equilibrium low-temperature plasma: IV. Ionisation and recombination functions, *Teplofizika Vysokikh Temperatur*, 7, No. 4, pp. 593–603.

131. Zhukov, M. F., Pustogarov, A. V., Dandaron, G.-N. B., *et al.* (1985) *Termochemical cathodes*, Novosibirsk, Institute of Thermophysics, 129 pp.

132. Beilis, I. I., Lyubimov, G. A. and Rakhovsky, V. I. (1969) Electric field on the electrode surface within the cathode spot of an arc discharge, *Dokl. Akad. Nauk SSSR*, **188**, pp. 552–555.

133. Moizhes, B. Ya. and Nemchinsky, V. A. (1972) To the theory of a high-pressure arc with refractory cathode: I., *Zh. Tekhn. Fiz.*, **42**, No. 5, pp. 1001–1009.

134. Moizhes, B. Ya. and Nemchinsky, V. A. (1973) To the theory of a high-pressure arc with refractory cathode: II., *Zh. Tekhn. Fiz.*, **43**, No. 11, pp. 2309–2317.

135. Zimin, A. M., Kozlov, N. P. and Khvesiuk, V. I. (1979) Calculation of the thermionic cathode, *Izv. Sib. Otd. Akad. Nauk SSSR, Ser. Tekhn. Nauk,* **2**, No. 8, pp. 17–24.

136. MacKeown, S. S. (1929) The cathode drop in an electric arc, *Phys. Rev.*, **34**, pp. 611–614.

137. Baksht, F. G., Dyuzhev, G. A., Mitrofanov, N. K., *et al.* (1973) Probe diagnostics of the low-temperature highly-ionised plasma, *Zh. Tekhn. Fiz.*, **43**, No. 12, pp. 2574–2583.

138. Dobretsov, L. N. and Gomoyunova, M. V. (1966) *Emission electronics*, Moscow, Nauka, 564 pp.

139. Fomenko, V. S. (1981) *Emission properties of materials.* A handbook, Kiev, Naukova Dumka, 338 pp.

140. Kikoin, I. K. (1976) *Tables of physical quantities.* A handbook, Moscow, Atomizdat, 1006 pp.

141. Noskov, M. M (1983) Optical and magneto-optical properties of metals, Sverdlovsk, UNC Akad. Nauk SSSR, 219 pp.

142. Kutateladze, S. S (1979) *Fundamentals of heat exchange theory*, Moscow, Atomizdat, 416 pp.

143. Krivtsun, I. V. and Som, A. I. (1999) Modeling of the laser-arc plasma torch, In: *Progress in Plasma Processing of Materials*, Proceedings of the 5th International Thermal Plasma Processes Conference, St. Petersburg, Russia, July 13–16, 1998, 958 pp.

144. Dilthey, U. (1996) Capabilities of welding processes and their automation potential with a view to the year 2000, In: *Proceedings of the IIW Asian Pacific Welding Congress*, Manukau City, New Zealand, 4-9 February 1996, pp. 65–81.

145. Present and future technologies of laser welding (1994) *Working group on future technologies of laser welding*, Japan Laser Processing Soc., March 1994, pp.240–251.

146. Shida, T., Hirokawa, N. and Fujikura, N. (1998) Welding of aluminium alloys by using high power CO_2-laser in combination with MIG arc, In: *Proceedings of the 6th Int. Conference on Welding and Melting by Electron and Laser Beams*, Toulon, France, 15-19 June 1998, pp. 399–406.

147. Seyffarth, P., Anders, B., Hoffmann J., et al (1997) *Vergleich unterschiedlicher verfahrensvarianten des kombinierten lichtbogen-laserstrahlschweissens*, FDS – Forschungszentrum des Deutschen Schiffbaus, Hamburg, Bericht Nr. 275/1997, 66 pp.

148. Krull, P. (2000) *Biegewechselfestigkeit WIG-, laser- und laser-WIG-geschweisster sowie randschichtverfestigter aluminium-legierungen*, Dissertation, Institut fuer Schweisstechnik, Technische Universitaet Braunschweig.

149. Dilthey, U., Lueder, F. and Wieschemann, A. (1998) Process-technical investigations on hybrid technology laser beam-arc welding, In: *Proceedings of the 6th Int. Conference on Welding and Melting by Electron and Laser Beams*, Toulon, France, 15-19 June 1998, pp. 417-424.

150. Seyffarth, P., Anders, B., and Hoffmann J. (1997) Combined laser beam arc welding — a high efficiency welding technology, In: *Proceedings of the Int. Conference on the Joining of Metals* [JOM-8], Helsingoer, Denmark, pp. 120–125.

151. Guidelines for the approval of CO_2-laser welding, Germanischer Lloyd, Hamburg, 1996.

152. Biffin, J. (1999) Plasma augmented laser welding, *International Sheet Metal Review*, Autumn 1999, pp. 26–30.

153. Bonss, S. (1999) Neues hybridschweissverfahren erweitert anwendungsspectrum von hochleistungs-diodenlasern, Fraunhofer-Institut fuer Werkstoff- und Strahltechnik, Dresden, Annual Report, pp. 46–47.

Index